ENVIRONMENTAL EFFECTS OF
ORGANIC AND INORGANIC CONTAMINANTS IN SEWAGE SLUDGE

Commission of the European Communities

ENVIRONMENTAL EFFECTS OF ORGANIC AND INORGANIC CONTAMINANTS IN SEWAGE SLUDGE

Proceedings of a Workshop held at Stevenage, May 25—26, 1982

Edited by

R. D. DAVIS

Water Research Centre, Stevenage, United Kingdom

G. HUCKER

Department of the Environment, London, United Kingdom

and

P. L'HERMITE

*Commission of the European Communities,
Directorate-General for Science, Research and Development, Brussels, Belgium*

D. REIDEL PUBLISHING COMPANY
DORDRECHT : HOLLAND / BOSTON : U.S.A.
LONDON : ENGLAND

Library of Congress Cataloging in Publication Data
Main entry under title:

Environmental effects of organic and inorganic contaminants in sewage
    sludge.

    At head of title: Commission of the European Communities.
    Held at the Water Research Centre, Stevenage, U.K.
    Includes index.
    1. Sewage sludge—Environmental aspects—Congresses. I.   Davis,
R. D., 1949-   . II. Hucker, T. W. G., 1922-   . III.   L'Hermite,
P. (Pierre), 1936-   . IV. Commission of the European Communities.
V.   Water Research Centre (Great Britain)
QH545.S493E54   1983   628.5   83-3237
ISBN 90-277-1586-6

QH
545
.S493E54
1983

Publication arrangements by
Commission of the European Communities
Directorate-General Information Market and Innovation, Luxembourg

EUR 8022
Copyright © 1983, ECSC, EEC, EAEC, Brussels and Luxembourg
(except pp. 91-112)

Published by D. Reidel Publishing Company
P.O. Box 17, 3300 AA Dordrecht, Holland

Sold and distributed in the U.S.A. and Canada
by Kluwer Boston Inc.,
190 Old Derby Street, Hingham, MA 02043, U.S.A.

In all other countries, sold and distributed
by Kluwer Academic Publishers Group,
P.O. Box 322, 3300 AH Dordrecht, Holland

D. Reidel Publishing Company is a member of the Kluwer Group

Printed in The Netherlands   # 9282826

C O N T E N T S

F O R E W O R D

This publication constitutes the Proceedings of two Seminars held at
the Water Research Centre, Stevenage (United Kingdom) on May 25-26, 1982,
under the auspices of the Commission of the European Communities, as part
of the Concerted Action COST 68 ter "Treatment of Sewage Sludge". The
Seminars were convened by Working Party 5 (Environmental Effects) of the
Concerted Action to examine current knowledge concerning organic micropollutants (Seminar I) and inorganic contaminants (Seminar II) in sewage
sludge utilised on agricultural land.

Continuing extension of sewage treatment in Europe is generating more
sewage sludge and hence putting more pressure on disposal outlets. Agricultural land is a principal disposal route for sewage sludge and has the
advantage that it involves the productive use of sludge to improve soil
conditions and supply nutrients for crop growth. At the same time it is
the route most sensitive to problems associated with organic and inorganic
contaminants which may occur in sludge in higher concentrations than in
soil. Considerable research effort is in progress within the Community
to investigate the effects of these contaminants and to ensure that soil
fertility, crops, livestock and the human foodchain are properly protected where sewage sludge is used on agricultural land. It was the aim
of the Seminars to provide a forum for the exchange of recent research
results and ideas on this subject.

<u>SESSION I - EFFECTS OF ORGANIC MICROPOLLUTANTS</u>

Occurrence, behaviour and fate of organic micropollutants
during waste water and sludge treatment processes

Effects arising from the presence of persistent organic
compounds in sludge

A Canadian perspective on toxic organics in sewage sludge

Toxic organic compounds in town waste materials: their origin,
concentration and turnover in waste composts, soils and plants

Identification of some organic micropollutants in urban sewage
sludges

Presentation of the analytical and sampling methods and of
results on organo-chlorines in soils improved with sewage
sludges and compost

Discussion

OCCURRENCE, BEHAVIOUR AND FATE OF ORGANIC MICROPOLLUTANTS

DURING WASTE WATER AND SLUDGE TREATMENT PROCESSES

J.N.Lester

Public Health and Water Resource Engineering Section
Civil Engineering Department
Imperial College of Science and Technology
London

SUMMARY

The principal areas of concern when disposing of sewage sludges contaminated with organic micropollutants are outlined and the characteristics
of the substances described. Some of the criteria which may be used to
identify these substances are presented and the principal chemical classes
listed. The sources of these substances and the routes by which they enter
waste water treatment works are reviewed. Their fate and behaviour during
conventional two stage waste water treatment is considered with emphasis
upon their removal and biodegradation. It would appear that the majority of
these materials are recalcitrant and liphophilic; they tend, therefore,
to be closely associated with the sewage solids and it would seem probable
that they are only biodegraded to a limited extent. As a consequence of
this behaviour they are concentrated in the sewage sludges produced. Their
behaviour during biological sludge treatment has been the subject of very
limited study and it is not possible to draw conclusions from this limited
information about the majority of these substances. The occurrence and range
of concentrations of organic micropollutants in sewage sludges is reviewed
as far as the available data permits. It is concluded that the principal
area of concern is human health.

INTRODUCTION

There exists mounting evidence that long term exposure to low concentrations of certain organic chemicals can be an important factor in the development and manifestation of some chronic diseases. It is further believed that between 80% and 90% of cancer cases are of environmental origin, and, therefore, the extent to which the population is exposed to environmental chemicals has become of concern (1). When attempting to identify the contaminants of sewage sludge which could be of concern, it is useful to consider those which cause concern when present in waters. Since the principal reasons for concern are not dissimilar, in the case of sewage sludge, they are,

i) human health, particularly when sewage sludge is disposed to agricultural land or the sea;

ii) adverse effects on biological sludge treatment processes, anaerobic and aerobic digestion;

iii) effects upon the environment, particularly agricultural land and the sea.

The degree of concern will be influenced by the toxicity, persistence and bioavailability of the various substances. The persistence, or resistance to chemical and biological degradation is an important feature of any contaminant of sewage sludge. These substances will be essentially chemically stable in the sewage matrix and not readily amenable to biodegradation during aerobic biological sewage treatment. Their concentration in the sewage sludge will be dependent upon their concentration in the influent raw sewage to the waste water treatment works and their affinity for settleable sewage and bacterial solids.

SUBSTANCES OF CONCERN

It is impossible to identify all the organic substances which could be of concern when present in sewage sludge. There are two principal reasons for this,

i) the extremely large, and ever increasing number of compounds which are potentially involved;

ii) the paucity of information about their adverse effects on the environment and in particular their chronic effects on humans.

It is possible to produce a list of toxic elements of concern because their numbers are finite and as a consequence their environmental effects and toxicology have been studied over an extensive period. However, because of the factors above, the state of knowledge about organics is very limited.

The United States production of organic chemicals has grown from 4.5 x 10^6 tonnes in 1943 to 64.0 x 10^6 tonnes in 1972 (2). The rate of growth has been exponential, increasing by approximately 9% per annum, resulting in production doubling every 8 years. Not only the quantity, but also the total number of chemicals produced is growing. During 1974 there were reported to be 12,000 chemical compounds in use in the United States (3). In the United Kingdom, it has been suggested that the number of chemicals currently available could be between 10,000 to 20,000 and increasing by approximately 1,000 per annum (4). It would appear to be inevitable, therefore, that a proportion of these chemicals will enter waste water treatment plants, either as the result of their use by the consumer, the discharge of industrial effluents, by surface run-off or by volatilization and subsequent atmospheric deposition.

Two lists of substances are widely accepted to include the principal compounds of concern, they are the European Economic Community list, both black and grey (5), and the United States Environmental Protection Agency Priority Pollutants list (6). The criteria used in the formulation of both lists are similar and it is, therefore,not surprising that many groups of substances are common to both. The USEPA list is however somewhat more detailed.

The EEC lists include substances or classes of substances which cause particular concern because they are known or suspected to adversely effect

        a) human health and welfare,

        b) aquatic life,

        c) waste water treatment processes.

It should be noted that "in selecting priorities some consideration must be given to the likelihood of harmful substances reaching a living organism, whether animal or man in sufficient concentration for a sufficient period for harm to occur" (7).

In the United States compounds with adverse effects on humans and the natural environment were included in the Priority Pollutants list based on the following criteria,

        a) frequency of occurrence of the compound in water,

        b) the availability of chemical standards for the quantification of the material of concern,

        c) the amount of production of the substance,

        d) chemical stability of the substance.

In general these lists include the following classes of organic compounds, polychlorinated biphenyls(PCB), phenols, chlorinated hydrocarbon solvents, polynuclear aromatic hydrocarbons (PAH), organophosphorus compounds, organotin compounds, phthalate esters, petroleum hydrocarbons and organochlorine insecticides.

In addition to the above concern has been expressed on occasion about surfactants, detergent builders, polybrominated biphenyls, and non-phosphorus fire retardents (including pentabromotoluene).

SOURCES AND OCCURRENCE

In areas where industrial effluents are discharged to sewage treatment works, many synthetic organic compounds may be present in the raw sewage. Pollutants commonly occurring under such circumstances include aliphatic and aromatic hydrocarbons, petroleum hydrocarbons, benzenes, phenols, PAH, halogenated (particularly chlorinated) aliphatic, alicyclic and aromatic compounds, organochlorine pesticides, PCB and phthalate esters (8-16). Table 1 includes a summary of the major sources of these pollutants. In addition to industrial effluents as sources of organic micropollutants in waste water, it is evident that domestic sewage contributes a low-level background concentration of trace organic contaminants to raw sewage (1, 35, 36) either through usage or as the result of the contamination of products during manufacture.

A significant proportion of organic contaminants carried by sewers derives from urban run-off, consisting of stormwater from roads, motorways and paved areas. A great variety of organic compounds may be found in urban run-off, including aliphatic and aromatic hydrocarbons, PAH, fatty acids, ketones, phthalate ester plasticisers and other more polar compounds (1,11,21, 22). Solvent-extractable organics are dominated by petroleum hydrocarbons which are considered to arise mainly from motor fuels and oils, tyres and bitumen in road surfaces (1,22). This source of organic contaminants has been shown to be responsible for a significant petroleum hydrocarbon load to the environment. Run-off may contain mg 1^{-1} concentrations of total hydrocarbons (21) and during heavy rainstorms, urban run-off may increase raw sewage PAH concentrations by up to 100-fold over dry weather conditions (11, 37). Leachates from solid waste dumps and landfill sites may contain a wide range of organic contaminants (38,39). While run-off from agricultural land contains significant concentrations of pesticides, although only a small

Table 1.   Major sources of organic contaminants in waste water

Chemical class	Sources	Reference
Aliphatic and aromatic hydro-carbons (inclu-ding benzenes and petroleum hydrocarbons).	Petrochemical industry wastes	17,18
	Heavy/fine chemicals industry wastes	19
	Industrial solvent wastes	19
	Plastics, resins, synthetic fibres rubbers and paints production	19
	Coke oven and coal gasification plant effluents	20
	Urban run-off	21,22
	Disposal of oil and lubricating wastes	17
Polynuclear aromatic hydrocarbons	Urban run-off	23
	Petrochemical industry wastes	24
	Various high temperature pyrolytic processes	11
	Bitumen production	25
	Electrolytic aluminium smelting	25
	Coal-tar coated distribution pipes	26
Halogenated aliphatic and aromatic hydrocarbons	Disinfection of water and waste water	3,27
	Heavy/fine chemicals industry wastes	28-31
	Industrial solvent wastes and dry cleaning wastes	28-30,32
	Plastics, resins, synthetic fibres, rubbers and paints production	28-32
	Heat-transfer agents	28,29,31
	Aerosol propellants	29,30
	Fumigants	29-31
Organochlorine pesticides	Agricultural run-off	33,34
	Domestic usage	35
	Pesticide production	36
	Carpet mothproofing	37-39
	Timber treatment	40
Polychlorinated biphenyls	Capacitor and transformer manufacture	41
	Disposal of hydraulic fluids and lubricants	42
	Waste carbonless copy paper recycling	42
	Heat transfer fluids	43
	Investment casting industries	41
	PCB production	44
Phthalate esters.	Plastics, resins, synthetic fibres, rubbers and paint production	32
	Heavy/fine chemicals industry wastes	32
	Synthetic polymer distribution pipes	1,45

proportion of the applied compounds may be lost in this manner (34,46,47).
Under some circumstances both these sources may contribute to the contamination of sewage sludge although not to the same degree as urban run-off.

Rainwater has on numerous occasions been shown to contain many organic
compounds including PAH, PCB, organochlorine insecticides, low molecular
weight chlorinated hydrocarbons, plasticisers and industrial solvents (1,8,
11,27,48,49). It has been demonstrated that the average concentration of total organochlorine insecticides in the rainfall of the U.K. is about 150 ng
1^{-1}, but ranges from 3 to 300 ng 1^{-1} (50). Organic contaminants reach the
atmosphere because of evaporation from sites on the earth's surface. The
principal mode of entry to the environment of many organic compounds of high
volatility such as industrial solvents is in fact by evaporative loss at sites of manufacture and use (27). Return to the earth's surface occurs either
by particulate fall-out, rain, or vapour phase deposition (27). This contribution may be combined with the urban run-off and enter waste water treatment plants with combined sewerage systems.

REMOVAL OF CONTAMINANTS DURING WASTE WATER TREATMENT

Many organic substances of concern are non-polar and hence have a very
low water solubility. Such compounds tend, therefore, to adsorb strongly on
suspended particulate matter (51). This suggests that mechanical separation processes such as sedimentation will achieve substantial removal of
these materials into the primary and secondary sludges. The limited information on the distribution of synthetic organic compounds in waste water
treatment processes confirms this.

Many synthetic organic compounds, because of their non-polar and hydro-
phobic nature not only adsorb onto particulate matter, but also partition
into non-polar fat and lipid material present in raw sewage. This component
of the raw sewage including mineral oils, greases, waxes and surfactants,
some of which in varying degrees are resistant to degradation, could potentially represent an important mechanism for the concentration and transport
of these materials. However, this is difficult to assess, since very little
research has been carried out in this area (48,52).

The limited information on the behaviour of organic micropollutants
during waste water treatment is summarised in table 2. Whilst the reported
removals exhibit a wide range, conventional two-stage sewage treatment would
appear to remove 70 to 90% of these substances present in the influent

Table 2. Removal of Organic Pollutants in Waste Water Treatment Processes.

Compound	Process Type	Removal Efficiency %	Reference
Polynuclear Aromatic Hydrocarbons (PAH)	Primary sedimentation percolating filter	50	53
	Primary sedimentation percolating filter/ activated sludge	86	53
	Primary sedimentation/ activated sludge	99	53
Polychlorinated Biphenyls (PCB)	Primary sedimentation/ percolating filter	86	54
	Primary sedimentation/ activated sludge	71	13
	Primary sedimentation with various secondary and tertiary treatment processes	66	55
	Primary sedimentation	51	56
	Primary sedimentation pilot plant	58	57
Dieldrin	Primary sedimentation/ trickling filter	≤95	38
	Primary sedimentation	51	56
	Primary sedimentation/ equalisation lagoon	99	37
	Activated sludge pilot plant	0-63	58
Lindane	Primary sedimentation/ trickling filter	44	59
p,p'-DDE	Primary sedimentation	37	56
	Activated sludge pilot plant	0-79	58
Aldrin	Activated sludge pilot plant	0-41	58
2,4-D Alkyl Esters	Activated sludge pilot plant	0-100	58

sewage. Where data is available the removals during the primary sedimenta-
tion(mechanical) stage of conventional sewage treatment appear to be in the
range of 40-60%.Removals during secondary biological treatment vary between
0 and 100% even for the same compound. The role which biodegradation may play
in the removal of organic micropollutants has not been clearly established.

If a particular organic contaminant is present in raw sewage and it is
not removed or only partially removed by primary sedimentation, then its
subsequent removal could be significantly influenced by its amenability to
biodegradation during aerobic biological sewage treatment. Solubility in
water appears to be one of the factors limiting the biodegradation of spa-
ringly soluble compounds such as PAH and PCB (48,60). The limiting factor
for biodegradation appears to be the organic/water interface which is the
major site of biodegradation of organic compounds (48,60).

PAH are biologically active molecules and are, therefore, able to enter
into biochemical reactions. There is also evidence of their biosynthesis
(61). There would appear to be no evidence for their biodegradation during secondary sewage treatment. However, bacteria have been isolated from
soil which are able to aerobically biodegrade PAH (62,63);such bacteria are
present in activated sludge. Significant biodegradation of the less chlorinated PCB in laboratory scale activated sludge simulations has been reported
(64), but it is questionable whether this would occur in full-scale conventional plants. Industrial waste waters containing the organophosphorus insecticide parathion, have been successfully treated using modified biological sewage treatment, extended retention time, high biological solids and
alkalinity control resulted in the complete biodegradation of parathion and
its principal metabolite p-nitrophenol (65). The treatment of waste waters
from the herbicide manufacturing industry has been reported after dilution
with domestic sewage and oxidation in an aerated lagoon (66).

It is improbable that most organic micropollutants are biodegraded to
a significant degree during conventional aerobic waste water treatment. Information about their possible inhibitory effects on the biological treatment process is apparently limited to 10 of the 114 USEPA priority pollutants, whilst information about effects on the sensitive nitrifying bacteria
is limited to phenol and 2,4-dichlorophenol (6). Removal is principally by adsorption onto biological solids followed by sedimentation to form a secondary
sludge. Thus when organic micropollutants are present in raw waste waters
both primary and secondary sludges will be contaminated with them.

SEWAGE SLUDGE TREATMENT

Following primary and secondary sedimentation, sewage sludges may be
treated further by a variety of processes including anaerobic or aerobic digestion and dewatering procedures such as vacuum or pressure filtration, pri-
or to disposal. These processes may have an effect upon the organic substances present in sludge. However, the behaviour of organic micropollutants in
these processes has been the subject of only a very limited number of investigations.

The effect of heat treatment of sewage sludge (to facilitate dewatering) on the distribution of PAH has been investigated. They were found to be
strongly associated with the sludge solids before and after heat treatment
(67,68). Studies conducted using a pilot plant (68) indicated that PAH were
not degraded or solubilised, but remained associated with the sludge solids,
regardless of treatment conditions. Studies undertaken at three full-scale
treatment plants (67) at separate locations confirmed that sludge type and
origin had no effect on PAH distribution during treatment.

The behaviour of PCB and organochlorine insecticides during the chemical conditioning and dewatering of sewage sludges has been investigated (52),
at two sewage treatment works employing different processes (vacuum filtration and pressure filtration) and conditioning agents (aluminium chlorohy-
drate and a polyelectrolyte). The recovery of both classes of compound were
in general closely correlated with the recovering of solids in both processes
and was independent of conditioning aid. PCB, DDE and Lindane were found to
be associated quantitatively with the sludge cakes produced from both vacuum
filter and pressure filtration plants, whilst dieldrin was found to be present in the filtrate liquor from both processes. It was suggested that this
was a consequence of the greater water solubility of dieldrin as opposed to
those of Arochlor 1260 and DDE, or the fact that, for most industrial and
household applications, dieldrin is marketed in a formulation of xylene and
anionic surfactants in order to increase its aqueous solubility (69). How--
ever, somewhat surprisingly, γ-HCH, which has an aqueous solubility approximately four times that of dieldrin did not display a similar tendency.

The sludge treatment process most frequently employed is anaerobic digestion. A laboratory-scale study of the breakdown of the organochlorine insecticides lindane, aldrin, dieldrin, endrin, heptachlor, heptachlor epoxide
and DDT during the anaerobic and aerobic digestion of sewage sludges has been
conducted by Hill and McCarty (70). Degradation of all compounds was found

to be more rapid under anaerobic than aerobic conditions, with the exception of heptachlor epoxide and dieldrin, which were found to be particularly persistent. In order to elucidate the mechanism of degradation (whether predominantly chemical or biological), several sludge digesters were 'poisoned' with high concentrations of cobaltous and mercuric salts. In all cases, the rate of degradation was slower in the poisoned units. Additional experiments with thick 'biological active' sludges produced inconclusive results; only lindane and aldrin exhibited any increase in degradation rate whilst results of other compounds were considered to be within the overall experimental error. The study concluded that the mechanism of degradation was primarily biological, but with a significant chemical contribution, and that under anaerobic conditions, the insecticides evaluated could be ranked in the following order of increasing persistence: lindane, heptachlor, endrin, DDT, aldrin, heptachlor epoxide and dieldrin. The application of these results to full-scale operation may be questionable, however, since concentrations of added insecticides (10-100 mg 1^{-1}) were considerably greater than those which might reasonably be expected to occur in sewage sludges.

The problems associated with attempting to predict the behaviour of organics, even within similar systems, are highlighted by the case of the detergent builder nitrilotriacetic acid (NTA). Where this compound is used as a detergent builder it is a major contaminant of primary sewage sludge. Upon the addition of NTA to laboratory scale anaerobic digesters, treating primary sludge, no biodegradation of NTA was observed over a period of 120 days (71), although the digesters contained a bacterial population capable of producing volatile fatty acids and converting them to methane in a manner entirely consistent with the normal operation of this process. However, when the feed to the digesters was changed to a 1:1 (vol/vol) mixture of primary sludge and surplus activated sludge, the latter being aerobically acclimated to NTA, the digesters were able after a period of 60 days to remove the influent NTA at a concentration of 20 mg 1^{-1}. An additional study of the behaviour of NTA during the anaerobic digestion of sewage sludge has confirmed this earlier observation (72). No biodegradation of NTA was observed during a period of 120 days although some removal did occur. This removal was unaffected by concentrations of sodium azide sufficient to inhibit methane production and it was concluded that the removal observed was partially due to adsorption onto sludge solids. Adverse effects of organic micropollutants on biological sludge treatment processes appears to be limited to the inhibition of anaerobic sludge digestion by various chlorinated hydrocarbons(73,74).

OCCURRENCE IN SLUDGE

Many different types of organic micropollutants have been detected in sewage sludges, including the 114 organic Priority Pollutants stipulated by the United States Environmental Protection Agency (75,76). Amongst the more frequently occurring compounds are aliphatic, alicyclic and aromatic hydrocarbons, PAH, PCB, organochlorine pesticides, various chlorinated hydrocarbons, phenols and alkyl phenols, benzenes and alkyl benzenes, phthalic acid esters, petroleum hydrocarbons, herbicides and organophosphorus flame retardants (8-16, 77).

Concentrations of PAH in sewage sludges at three sewage treatment works in W.Germany were found to vary between 1.6 and 6.0 mg kg^{-1}(78).Analysis of sewage sludge from a sewage treatment works in the UK receiving predominantly domestic sewage revealed PAH concentrations of 0.8-7.0 µg l^{-1} (67). A limited survey of sewage sludges from 12 UK sewage treatment works found concentrations of individual PAH varying between 0.08 and 11.4 µg g^{-1}, with the highest concentrations present in sludges from those treatment works serving industrialised areas (79). Profile analysis of sewage sludge samples from a treatment works in Germany indicated that concentrations of PAH were present in the range 0.09 to 4.39 µg g^{-1}; a total of 18 PAH were identified (80).

A study of 22 sewage works in Canada revealed concentrations of PCB in the primary and digested sludges varied between 0.4 and 38.4 µg g^{-1} (55). A detailed study of PCB in one US sewage treatment plant reported an average PCB content in the digested sludges of 765.0 µg g^{-1} (54).A recent survey of 40 sewage sludges in the UK has reported concentrations of PCB which varied between 0.02 and 0.46 µg g^{-1}, with a mean of 0.16 µg g^{-1} (81). Two previous analytical surveys had reported PCB concentrations in 10 sewage sludges of between 0.004 µg g^{-1} and 2.680 µg g^{-1} (82,83).

Analysis for organochlorine pesticides in 40 UK sewage sludges revealed that concentrations of three compounds, lindane, DDE and dieldrin, varied between 0.01 and 1.26 µg g^{-1} (81). In the US significant concentrations of certain insecticides, including DDT, DDD, dieldrin, aldrin and chlordane, of between 0.31 and 16.20 µg g^{-1} occurred in sewage sludges (84) and other chlorinated organic compounds have also been detected in US sewage sludges (14). Among the less persistent organic compounds detected in sludges are organophosphorus and organobromine flame retardant chemicals (85,86), some of which are toxic (87). These concentrations are summarised in table 3.

Table 3. Occurrence of Organic Micropollutants in Sewage Sludges

Compounds	Sample Type (country)	Concentration	Reference
Polynuclear Aromatic Hydrocarbons (PAH)	Primary & digested sludges (Germany)	1.6-6.0 mg kg^{-1}	78
	Primary sludges Activated sludges (UK)	2.6-10.0μg l^{-1} 0.8-7.0μg l^{-1}	67
	Primary sludge heat-treated sludge (UK)	0.1-10.0 mg kg^{-1} 0.09-16.0mg kg^{-1}	68
	12 sludges primary/digested	0.08-11.4mg kg^{-1}	79
	Primary sludges (Germany)	0.09-4.39mg kg^{-1}	80
Polychlorinated Biphenyls (PCB)	22 primary and digested sludges (Canada)	0.4-38.4mg kg^{-1}	55
	Digested sludge (USA)	765.0 mg kg^{-1}	54
	40 primary and digested sludges (UK)	0.02-0.46mg kg^{-1}	81
	10 sludges, primary and digested	0.004-2.680mg k^{-1}	82,83
	16 sludges digested (USA)	0.01-23.1 mg kg^{-1}	88
Lindane	40 primary and digested sludges (UK)	<0.01-0.93mg kg^{-1}	81
Dieldrin	40 primary and digested sludges	<0.01-1.26mg kg^{-1}	81
	16 digested sludges (USA)	<0.01-2.2mg kg^{-1}	88
DDE	40 primary and digested sludges (UK)	0.01-0.49mg kg^{-1}	
Organophosphorus	3 primary sludges (UK)	750-6000μg l^{-1}	85
Pentabromotoluene	4 primary sludges (Sweden)	8.0-180.0mg kg^{-1}	86

CONCLUSIONS

The principal concern in disposing of sewage sludges contaminated with organic micropollutants is the same as that associated with the discharge of sewage effluents contaminated with these substances, that is the exposure of the population to these substances either directly from the consumption of water or indirectly through the food chain. When sludge is disposed to agricultural land the contaminants present may be taken up by crop plants or ingested by grazing animals, they may also enter surface waters as the result of run-off or penetrate ground waters. When disposed to sea they may be introduced into potential human food chains, where biomagnification may occur.

Information dealing with the toxicity of organic micropollutants to biological waste water and sludge treatment processes is insufficient to permit the magnitude of this potential problem to be evaluated. In addition, data describing the biodegradation of these substances during waste water and sludge treatment are inadequate and it is not, therefore, possible to determine the contribution this process may make to the destruction of these substances.

REFERENCES

1. M.Fielding and R.F.Packham, J.Inst.Water Eng.Sci., 31 (1977) 353-375.

2. R.L.Metcalf. In: I.H.Suffet (Ed.),Fate of Pollutants in the Air and Water Environments, Part 2, John Wiley and Sons, New York, USA, 1977, pp.195-221.

3. M.T.Gillies (Ed.), Drinking Water Detoxification, Noyes Data Corporation, Park Ridge, New Jersey, USA, 1978.

4. A.Waggott and A.B.Wheatland. In:O.Hutzinger, I.V.Van Lelyveld and B.C.J.Zoeteman (Eds.), Aquatic Pollutants: Transformation and Biological Effects, Pergamon Press, Oxford, UK, 1978, pp.141-168.

5. European Economic Community. Council on pollution caused by certain dangerous substances discharged into the aquatic environment of the community (76/464/EEC-OJL). Off.J.Eur.Commun. L 129.

6. R.M.Athony and L.H.Breimhurst. J.Water Pollut.Control Fed., 53 (1981) 1457-1468.

7. Water Research Centre. Emission standards in relation to water quality of objectives. TR 17, Medmenham, UK 72pp.

8. W.Giger, M.Reinhard, C.Schaffner and F.Zürcher, In: L.H.Keith (Ed.), Identification and Analysis of Organic Pollutants in Water, Ann Arbor Science, Ann Arbor, Michigan, USA, 1976, pp.433-452.

9. A.W.Garrison, J.D.Pope and F.R.Allen. In:L.H.Keith (Ed.), Identification and Analysis of Organic Pollutants in Water, Ann Arbor Science, Ann Arbor, Michigan, USA, 1976, pp.517-556.

10.   A.L.Burlingame, B.J.Kimble, E.S.Scott, F.C.Walls, J.W.de Leeuw,
      B.W.de Lappe and R.W.Risebrough. In:L.H.Keith (Ed.), Identification
      and Analysis of Organic Pollutants in Water, Ann Arbor Science, Ann
      Arbor, Michigan, USA, 1976, pp.557-585.

11.   R.M.Harrison, R.Perry and R.A.Wellings, Water Res., 9 (1975) 331-346.

12.   W.W.Pitt, Jr., R.L.Jolley and C.D.Scott. Environ.Sci.Technol., 9,
      (1975) 1068-1073.

13.   D.J.Dube, G.D.Veith and G.F.Lee. J.Water Pollut.Control Fed., 46,
      966-972.

14.   M.D.Erickson and E.D.Pellizzati. Bull.Environ.Contam.Toxicol., 22,
      (1979) 688-694.

15.   J.Lawrence and H.M.Tosine. Bull.Environ.Contam.Toxicol., 17 (1977)
      49-56.

16.   H.E.Wise, Jr., and P.D.Fahrenthold.Environ.Sci.Technol., 15(1981)
      1292-1304.

17.   D.W.Connell and G.J.Miller. CRC Crit.Rev.Environ.Control.,11 (1981)
      37-162.

18.   A.A.Wigren and F.L.Burton.J.Water Pollut.Control Fed., 44 (1972)
      117-128.

19.   L.Fishbein. Sci.Total Environ., 17 (1981) 97-110.

20.   C.W.Fisher. In: F.F.Gurnham (Ed.), Chemical Technology, Vol.2: Indus-
      trial Wastewater Control, Academic Press, New York, USA, 1965.

21.   J.V.Hunter, T.Sabatino, R.Gomperts and M.J.MacKenzie. J.Water Pollut.
      Control Fed., 51 (1979) 2129-2138.

22.   R.P.Eganhouse, B.R.T.Simoneit and I.R.Kaplan. Environ.Sci.Technol.,
      15 (1981) 315-326.

23.   M.Waibel. Zbl.Bakt.Hyg., I.Orig.Reike B., 163 (1976) 458-469.

24.   I.A.Veldre, L.A.Lakhe and I.K.Arro. Hyg.Sanit., 30 (1965) 291-294.

25.   K.P.Ershova. Hyg.Sanit., 33 (1968) 268-270.

26.   Water Research Centre, A Survey of Polycyclic Aromatic Hydrocarbon
      Levels in British Waters, TR 158, Medmenham, UK, 1981, 47pp.

27.   G.McConnell. J.Inst.Water Eng.Sci., 30 (1976) 431-445.

28.   L.Fishbein. Sci.Total Environ., 11 (1979) 111-161.

29.   L.Fishbein. Sci.Total Environ., 11 (1979) 163-195.

30.   L.Fishbein. Sci.Total Environ., 11 (1979) 223-257.

31.   L.Fishbein. Sci.Total Environ., 11 (1979) 259-278.

32.   L.Fishbein and W.G.Flamm. Sci.Total Environ., 1 (1972) 117-140.

33.   F.Moriarty. In: F.Moriarty (Ed.), Organochlorine Insecticides: Persi-
      stent Organic Pollutants. Academic Press, London, UK, 1975, pp.29-72.

34.   M.G.Browman and G.Chester. In: I.H.Suffet (Ed.), Fate of Pollutants
      in the Air and Water Environments, Part 1, John Wiley and Sons, New
      York, USA, 1977, pp.49-105.

35.   G.T.Brooks. In:F.Matsumura, G.M.Boush and T.Misato (Ed.), Environmen-
      tal Toxicology of Pesticides, Academic Press, London, UK, 1972,
      pp. 61-113.

36. Environmental Protection Agency, Development Document for Interim Final Effluent Limitations - Guidelines for the Pesticide Chemical Manufacturing Point-Soruce Category, Document No.EPA 440/1-75/060d, Washington, DC, USA, 1976, 381 pp.

37. L.Brown, E.G.Bellinger and J.P.Day. J.Inst.Water Eng.Sci.,33, (1979) 478-484.

38. R.D.Wilroy. Proc. 18th Ind.Waste Conf., 115 (1963) 413-417.

39. W.F.Bernholz. J.Barandy and N.Kins. Proc.26th Ind.Waste Conf., 140 (1971) 65-76.

40. Department of the Environment, The Non-Agricultural Uses of Pestici-des in Great Britain, H.M.S.O.,London, UK, 1974, 65pp.

41. J.L.Hesse, Polychlorinated biphenyl usage and sources of loss to the environment in Michigan. Proc.Nat.Conf. on PCB's, November 19-21, 1975, Chicago, Illinois, Environmental Protection Agency Publication No. EPA 560/6-75-004, Washington, DC, USA, 1976, pp.127-133.

42. S.J.Kleinert, Sources of polychlorinated biphenyls in Wisconsin.Proc. Nat.Conf. on PCB's, November 19-21, 1975, Chicago, Illinois, Environ-mental Protection Agency Publication No.EPA 560/6-75-004, Washington, DC, USA, 1976, pp.124-126.

43. L.Fishbein. Sci.Total Environ., 2 (1973) 305-340.

44. Environmental Protection Agency, PCB's in the United States: Indus-trial Use and Environmental Distribution, Report No.EPA 560/6-76-005, Washington, DC, USA, 1976, 334pp.

45. G.A.Junk, H.J.Svec, R.D.Vick and M.J.Avery. Environ.Sci.Technol., 8, (1974) 1100-1106.

46. L.E.Asmussen, A.W.White, Jr., E.W.Hauser and J.M.Sheridan. J.Environ. Qual., 6 (1977) 159-162.

47. C.A.Edwards, In: C.A.Edwards (Ed.), Environmental Pollution by Pesti-cides, Plenum Press,London, UK, 1974, pp.409-458.

48. R.R.Lock, A Review of the Sources, Behaviour and Fate of Polynuclear Aromatic Hydrocarbons, Polychlorinated Biphenyls and Insecticides in Waste Water Treatment Processes, M.Sc.Thesis, University of London, UK, 1978, 137 pp.

49. W.A.Hoffman, Jr., S.E.Lindberg and R.R.Turner. Environ.Sci.Technol., 14 (1980) 999-1002.

50. R.L.Jolley. J.Water Pollut.Control Fed., 47 (1975) 601-618.

51. S.E.Herbes. Water Res., 11 (1977) 413.

52. A.E.McIntyre, J.N.Lester, R.Perry. Environ.Pollut. (Series B) 2 (1981) 309-320.

53. J.Reichert, H.Kunte,K.Engelhardt and J.Borneff, Zbl.Bakt.Hyg., 155, (1971) 18-40.

54. A.K.Bergh and R.S.Peoples, Sci.Total Environ.,8 (1977) 197-204.

55. E.E.Shannon, F.J.Ludwig and I.Valdemanis, Polychlorinated Biphenyls in Municipal Water Waters: an assessment of the problem in the Canadian Lower Great Lakes. Environment Canada, Report No.73-3-8, (1976) 35pp.

56. A.E.McIntyre, R.Perry and J.N.Lester, Environ.Pollut. (Series B) 2, (198)

57. A.Garcia-Guiterrez, A.E.McIntyre, R.Perry and J.N.Lester, Sci.Total Environ., 22 (1982) 243-252.

58. F.Y.Saleh, G.F.Lee and H.W.Wolf, J.Wat.Pollut.Control Fed., 52 (1980) 19-28.

59. D.B.Harper, R.V.Smith and D.M.Gotto, Environ.Pollut.,12(1977)223-233.

60. D.L.S.Liu, Environ.Conserv.,3 (1976) 137-138.

61. J.M.Neff. Polycyclic Aromatic Hydrocarbons in the Aquatic Environment. Applied Science Publishers, Barking, UK.

62. J.I.Davies and W.C.Evans, Biochem.J., 91 (1964) 251-261.

63. R.S.Wodinski and M.J.Johnson. Appl.Microbiol.,16 (1968) 1886-1991.

64. E.S.Tucker, V.W.Saeger and O.Hicks, Bull.Environ.Contam.Toxicol., 14 (1975) 705-713.

65. G.Coley and C.N.Stutz. J.Water Pollut.Control Fed.,38(1966)1345-1349.

66. Environmental Protection Agency, Biological Treatment of Chlorophenolic Wastes, Water Pollution Research Series, Report No.EPA-12130-Egk 06/81, Washington, DC, USA, 1971, 187pp.

67. T.P.Nicholls, R.Perry and J.N.Lester. Sci.Total Environ.,12,(1979) 137-150.

68. T.P.Nicholls,J.N.Lester and R.Perry.Sci.Total Environ.,14(1980)19-30.

69. L.Brown,E.G.Bellinger and J.P.Day. J.Inst.Water Eng.Sci.,33 (1979) 478-484.

70. D.W.Hill and P.L.McCarty.J.Water Pollut.Control Fed.,39(1967)1259-1277.

71. L.Moore and E.F.Barth. J.Water Pollut.Control Fed.,48(1976)2406-2409.

72. P.W.W.Kirk,J.N.Lester and R.Perry. Water Res., 16 (1982) in press.

73. J.D.Swanwick and M.Foulkes. Water Pollut.Control., 69 (1970) 58-70.

74. D.P.Stickley.Water Pollut.Control., 69 (1970) 585-592.

75. L.H.Keith and W.A.Telliard. Environ.Sci.Technol.,13 (1979) 416-423.

76. D.F.Bishop. EPA Environ.Res.Lab.,Cincinnati, Ohio,USA,1980, 56pp.

77. W.A.Telliard. EPA. Effluent Guidelines Div.USA, 1977, 35pp.

78. J.Borneff and H.Kunter. Arch.Hyg.Bakt., 149 (1965) 226-243.

79. A.E.McIntyre,R.Perry and J.N.Lester.Anal.Lett., 14 (1981) 291-309.

80. G.Grimmer, H.Bohnke and H.Browitzky.Fresenius'Zeitschr.Anal.Chem., 289 (1978) 91-95.

81. A.E.McIntyre and J.N.Lester.Environ.Pollut.(Series B)3(1982)225-230.

82. Water Research Centre.PCB in Sewage Sludges 1971-72,UK,1973, 32pp.

83. Water Research Centre.PCB in Sewage Sludges 1972-73,UK, 1974, 37pp.

84. J.G.Babish et al. Spec.Rept.No.42, Cornell Univ.NY, USA, 1981, 5pp.

85. A.E.McIntyre, R.Perry and J.N.Lester. Bull.Environ.Contam.Toxicol., 26 (1981) 116-123.

86. P.E.Mattsson et al. J.Chromatog., 111 (1975) 209-213.

87. L.Fishbein, Rept. PB-273-229, NTIS,Springfield, Virg.,USA,1977, 41pp.

88. A.K.Furr et al. Environ.Sci.Technol., 10 (1976), 683-687.

EFFECTS ARISING FROM THE PRESENCE OF
PERSISTENT ORGANIC COMPOUNDS IN SLUDGE

D. G. LINDSAY

Food Science Division, Ministry of Agriculture, Fisheries and Food

Great Westminster House, Horseferry Road, LONDON SW1P 2AE

Summary

Disposal of sewage sludge containing residues of persistent organic compounds can result in contamination of the food chain and consequential human exposure.   The disposal of contaminated sludges to sea has resulted in increased residue levels in fish. However any effects are likely to be small in comparison with land disposal due to more effective sea dispersion. The greatest effects are likely to be from the disposal of contaminated sewage sludge to pasture land which can result in the contamination of meat and dairy products.   It is estimated that residues in milk on a "worst case" assumption could reach 0.5ppm for every 1.0ppm of the persistent chemical present in sludge dry matter if a 30 day no-grazing interval after disposal is observed.   In line with the general policy of avoiding contamination of food by such chemicals wherever practicable, sludges known to contain such residues should not be applied to permanent pastures.

Introduction

The primary degrading organisms in activated sludge treatment plants are
bacteria which derive their nutritional requirements from compounds present
in the influent waste.   Energy is derived principally from the hydrolysis
and oxidation of the organic compounds present in the sewage and the energy
released through oxidative phosphorylation of the intermediates of the
tricarboxylic acid cycle in the form of ATP.   This energy is used to
support growth of the bacterial population in the activated sludge process.

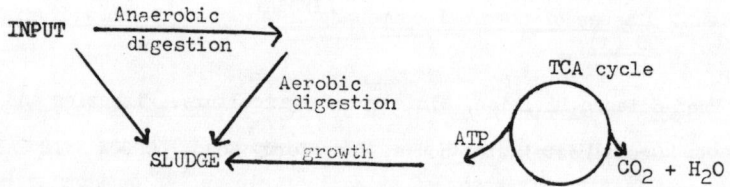

Fig 1. The activated sewage sludge process

Thus, unlike the inorganic constituents of sewage, where bacterial action
does not result in the conversion of the compounds into simple volatile
and soluble compounds, organic constituents rarely survive the sewage
treatment processes.

Degradation of organic compounds in soils and the aquatic environment

Should the contact time in the digesters be insufficient for complete degradation of the chemical, further degradation of any complex organic
compounds will take place through bacterial and chemical reactions on disposal.   Many of the bacteria present in soils and sediments are those
which are utilised in the sludge treatment process itself and thus few
compounds are likely to persist for more than a few weeks at the most from
the period of disposal of sludge.

Persistent organic compounds in sludge

Nonetheless there have been numerous illustrations in the recent past where
organic compounds have been shown to have a fairly high persistence in the
environment.   The previous speaker has described some of the classes of
chemicals which survive the activated sludge process.   All of these compounds have the potential to survive for long periods in the environment
and some of the compounds which have been detected in freshwater fish in
the US and are known to survive the sewage treatment process are shown in
Table 1.

Aromatic amines
Brominated aromatics
Chlorinated aliphatics
Chlorinated benzenes
Chlorinated benzo fluorides
Chlorinated non-aromatic cyclics
Chlorinated toluenes
Triaryl phosphates

Table 1: <u>Classes of chemical compounds detected in freshwater fish in the US (1)</u>

The aerobic breakdown of chemicals by micro-organisms is largely confined to compounds which are able to pass into solution within the soil. If a compounds is strongly absorbed by soil or sludge, as appears to be the case for highly persistent organic chemicals, the rate of disappearance of the compound may be largely limited to chemical processes and be significantly slower. It is also likely that enzymes causing the breakdown of the chemical have not been developed by organisms to metabolise, at any significant rate, those chemicals which have no structural similarities to naturally occurring chemicals.

Metabolism or organic chemicals by sludge bacteria

Although individual classes of chemical may be easily metabolised by sludge bacteria, it does not follow that this will lead to progressive oxidative metabolism. Thus a 50% degradation of tri(2,3-dibromopropyl) phosphate was observed in a laboratory activated sludge system in 24 hours. However the compound was not completely degraded and the product formed was found to be bis (2,3-dibromopropyl) phosphate which resists further breakdown(2).

CONTAMINATION OF FOOD BY ORGANIC CHEMICALS IN SLUDGE

There are four possible ways in which organic compounds can find their way into the food chain as a result of sludge disposal.

They are:
1. the direct contamination of food crops by spraying;
2. the uptake and translocation of such compounds from soil into the growing crop;
3. the ingestion of soil and contaminated pasture by grazing animals; and
4. the accumulation in fish and shellfish as a result of the disposal of sewage sludge at sea.

Guidelines for the disposal of sludge recommend that sludges should not be sprayed onto growing crops to be used directly as human food and thus the first route is likely to be an insignificant source of residues in the diet.

As explained earlier the most persistent organic chemicals appear to be tightly bound to soils. Studies which have been carried out on the translocation of pesticides from soils into growing crops have shown that persistent organochlorine pesticides and the polyhalogenated biphenyls are almost totally unavailable to the plant and are thought to be tightly bound to the humic acid fraction (3,4,5). Volatilisation of these compounds from the soil is possible but this is likely to be limited to the soil surface and compounds present in the soil profile are likely to be an insignificant source of residues in the growing crop. Although there has been a report of the translocation of persistent organochlorine compounds from soil into oil-seed crops (6), it is probable that this is an insignificant problem in the UK as a source of residues in food even if these results were to be confirmed by other workers.

The third route is potentially the most important source for contamination of the food chain. Sludges are known to adhere to herbage and ingestion of pasture and contaminated soil by animals can result in the accumulation of residues in meat and dairy products. The systems developed by mammals to detoxify ingested chemicals are no more capable of degrading and eliminating persistent chemicals than are soil or sludge bacteria. Since most of the persistent chemicals are also strongly lip ophilic, they are stored in the animal's fat cells. During period of food deprivation these energy stores are mobilised and the chemical released into the blood stream only to be reincorporated into fat storage cells or eliminated through the excretory organs. Eventually a steady state arises in which a positive balance is achieved between storage and excretion. However in the case of the lactating animal, fat is released directly into the milk and with it, the associated persistent chemical residue.

The average amount of soil ingested by cattle has been estimated by Healy to be 6% of the dry matter intake of the animals (7). During the period when the pasture growth was at a minimum the monthly average soil intake was 14% of the dry matter intake. These calculations were made from experiments on cattle which did not receive supplemental feeds. Although sheep can also consume significant amounts of soil, especially in winter, the annual average intakes were found to be no higher than cattle.

Experiments undertaken by Chaney and Lloyd (8) have shown that for sludge
applied at the maximum practicable rate of application to pasture and to
which cattle were given immediate access, 30% of the forage dry matter
intake consisted of sludge. After a no-grazing interval of 30 days the
sludge intake declined to 10% of forage dry matter. Equilibrium milk fat
concentrations of some organochlorine pesticides and PCB's were found to
be 5 times the concentrations of PCB's in forage dry matter (8).
On the basis of these results a "worst case" estimate for the concentration
of a persistent residue in milk fat from soil would be 0.3 ppm for every
1ppm of the chemical residue present in soil (1 x 0.06 x 5).   The concentration of a residue in milk derived from sludge ingestion would be 0.5
ppm for every 1ppm of the residue present in sludge if a no-grazing interval
of 30 days were observed (1 x 0.1 x 5).   These calculations are worst-
case assumptions since most cattle receive feed supplements, particularly
in the winter months.   The actual amounts of soil or sludge ingested
obviously depends on many factors and equilibrium concentrations of the
persistent chemical in animal tissues will be reached only after extensive
exposure.

Evidence implicating the disposal of sewage sludge on pasture land as a
cause of contaminated milk has been obtained in the case of dieldrin
when above average levels of dieldrin were detected in milk sampled from
farms which had received contaminated sludges (11,12).

In addition to the contamination of meat and dairy products, fish can
accumulate residues from the discharge of contaminated sludges at sea.
Fish will feed off suspended solids, to which the chemicals are often
strongly bound, and species with a high oil content such as herring will
accumulate the chemical if disposal is not into waters where dispersion is
rapid.   Higher than average levels of PCBs and dieldrin have been detected in the past in fish caught off the Garroch Head in the Firth of Clyde
where sludge dumping occurs (13,14).   The inputs of these chemicals into
the sewage treatment plant has subsequently been controlled and residue
levels have fallen in consequence.   However disposal to the marine
environment is likely to present less of a problem than disposal to land
because of the rapid dispersion and dilution which normally occurs.

EFFECTS ON HUMAN HEALTH
The contamination of food and drinking water by industrial chemicals is the
major source of exposure of the general population to environment contaminants for all except the most volatile chemicals.   Potential effects arising

from the contamination of food may arise from the gradual diffusion of
persistent chemicals throughout the environment resulting in a low-level
and long-term exposure to a major proportion of the population.

A major concern arises from the possibility that health effects could arise
from the presence of residues of chemicals in the food supply which are
known, or suspected, to be human carcinogens. Studies in experimental
animals have supported the view that for some carcinogens it is difficult
to determine a dose at which there is no effect. Increasing the size of
the test group of animals, or choosing a different genetic strain can lead
to an observable increase in the incidence of tumours. These observations
create insuperable difficulties in determining an acceptable level of
human exposure which has resulted in the adoption of rigorous controls.

Apart from the potential health risks associated with environmental
chemicals which are carcinogenic, potential chronic health effects may
also arise from the accumulation of these toxic substances in human tissues.
The factor which cause these chemicals to accumulate in animal, fish or
plant tissues are also responsible for accumulation in man. Organic
chemicals can accumulate in tissues and in some instances cross usually
effective barriers such as the placenta or blood-brain barrier through
transport associated with plasma proteins or binding to erythrocytes, and
may concentrate in excretory and other organs. Thus potential long-term
health effects may include birth defects and reproductive disorders in
addition to systemic toxicity. Under conditions of stress or unusual
metabolic activity resulting in the mobilisation of fat stores, high
levels of the substance may be liberated into the blood stream and into
breast milk resulting in levels of exposure to the neonate which are
greater than those which would normally be encountered in infant foods.

The concern for infants is based on the fact that they consume a very
restricted diet and are known to be more susceptible to the effects of
certain toxins than are adults.

All of these factors which make risk-assessment of trace contaminants of
the diet so difficult to undertake, have meant that controls over sources
of contamination of stable foods have been applied even in the absence of
definite evidence of harm to human health. A disposal policy is sought
which attempts to minimise the contamination of the food chain through an

awareness of sources and pathways of entry of persistent organic chemicals into the food chain. In the case of sludge disposal this is best achieved if sludge which is known to contain residues of persistent organic compounds is preferably not disposed of to permanent pasture land.

References

1.   Office of Technology Assessment (1979). "Environmental Contaminants in Food" Congress of the United States. Washington. p83.

2.   Alvarez, G.H., Page, S.W., Ku, Y. 1982. Biodegradation of ^{14}C-tris (2,3-dibromopropyl) phosphate in a laboratory activated sludge system. Bull. Env. Cont. Toxicol. 28 : 85-90.

3.   Beall, M.L and Nash. R.G. 1971 Organochlorine insecticide residues in soybean plant tops : root vs. vapor sorption. Agron.J. 63 : 460-464.

4.   Jacobs, L.W., Chou, S.F. and Tiedje, J.M. (1976). Fate of polychlorinated biphenyls in soils. Persistence and plant uptake. J.Agric. Food Chem 24 : 1198-1201.

5.   Fries, G.F. and Marrow G.S. 1981 Chlorobiphenyl movement from soil to soybean plants. J.Agric. Food Chem 29 : 757-759.

6.   Moore S., Petty, H.B. and Bruce W.N. (1976) Insecticide residues in soybeans in Illinois 1965-74. 28th Illinois Custom Spray Operators Training School. Summary of Presentations, 226-230.

7.   Healy, W.B. 1968. Ingestion of soil by dairy cows. N.Z.J. Agric. Research 13 : 664-672.

8.   Chaney, R.L., and Lloyd, C.A. 1979. Adherence of spray-applied liquid digested sewage sludge to tall fescue. J. Environ. Qual. 8 : 407-411.

9.   Fries, G.F. (1982) Potential polychlorinated biphenyl residues in animal products from application of contaminated sewage sludge to land. J. Environ. Qual., 11, 14-20.

10. Braund D.G., Brown, L.D., Huber J.T., Leeling, N.C and Zabik, M.J. (1969). Excretion and storage of dieldrin in dairy cows fed thyroprotein and different levels of energy. J. Dairy Sci., 52, 172.

11. Lindsay, D.G. (1979) Possible health hazards from the presence of persistent chemical residues in sewage sludge. Conference on Utilisation of Sewage Sludge on Land. Water Research Centre. C.E.C. pp 238-245.

12. DeHaan, T.A.M. (1976) Proc. Int. Conference on Renovating and Reuse of Wastewaters through aquatic and terrestial systems. Bellagio, Italy, 1975. Marcel Decca. New York.

13.   Gattney, P.E. (1976).   Carpet and rug industry case study 1 : water and wastewater treatment plant operation.   J. Water Polln.   Contr. Fed., 48, 2590.

14.   Waddington, J.I., Best, G.A., Dawson, J.P., and Lithgow. T. (1973) PCB's in the Firth of Clyde.   Marine Polln. Bull. 4, 26-28.

A CANADIAN PERSPECTIVE ON TOXIC ORGANICS IN SEWAGE SLUDGE

T.R. BRIDLE AND M.D. WEBBER

Wastewater Technology Centre
Environmental Protection Service
Environment Canada
Burlington, Ontario, L7R 4A6, Canada

Summary

Analytical methods for toxic organic compounds in sewage sludge are not highly refined, however, experience with the methods and confidence in the results is increasing. Evidence is presented for the occurrence of a wide variety of toxic organic compounds including polynuclear aromatics, phthalates and chlorinated hydrocarbons in Canadian and USA sewage sludges.

Toxic organic compounds should be considered when deciding upon sludge processing, utilization and disposal options. These compounds generally exhibit low water solubility, are bioresistant and accumulate in the environment, and translocate up the food chain. However, more specific information concerning their fate during processing, land application and incineration of sludge is very limited. The Wastewater Technology Centre has recently initiated research in each of these areas of concern. It should provide much needed information on which to base environmentally sound decisions.

1. INTRODUCTION

Large numbers of organic compounds of environmental concern have been identified in both Canada and the USA. In Canada, the Water Pollution Control Directorate of Environment Canada has developed a "WPCD Working List of Suspect/Priority Toxic Chemicals" (1). The list is comprised of 150 compounds or groups of compounds and was derived from existing Health and Welfare Canada, Environment Canada, Great Lakes Water Quality Board and USEPA Priority Pollutant lists. In the USA, the Environmental Protection Agency has identified 129 "Priority Pollutants" including 114 organics, 13 metals, cyanide and asbestos (2).

Toxic organic compounds enter the sewerage system from domestic and industrial sources and may accumulate in the sludge. They are potentially hazardous to humans and animals for five principal reasons (3).

1. They have very low water solubility and do not move readily in soil.
2. They are relatively stable in soil because they are resistant to microbial degradation.
3. They are fat soluble and bioaccumulate in tissue.
4. They accumulate and translocate up the food chain (soil to plants to animals to man).
5. They are highly toxic to mammals; many of them are carcinogenic, mutagenic and teratogenic.

Consequently, the nature and quantity of toxic organics in sludge should be considered prior to deciding upon a utilization or disposal technique. The purpose of this paper is to; (a) discuss sampling and analysis of sludge for toxic organic compounds, (b) present evidence for toxic organics in Canadian and USA sludges and, (c) discuss the Wastewater Technology Centre (WTC), Burlington research program on toxic organics in sludge.

2. SAMPLING AND ANALYSIS OF SLUDGES FOR TOXIC ORGANICS

Wastewater treatment plant sludges are heterogeneous and it is difficult to obtain representative samples for analysis (4). Sampling for analysis of toxic organics poses the same problems as for conventional parameters. Precautions should be taken to ensure that representative samples have been obtained. In addition, samples for toxic organic analysis should be collected in glass bottles with teflon or aluminum lined tops to avoid contamination.

Even when representative samples have been obtained, significant problems still exist with the analysis of toxic organics in sludges (5). Major studies are underway in Canada and the USA to refine the analytical protocols (6,7). The need for refinement is evident since recoveries of known compounds added to sludge are erratic. Petrasek et al (8) reported that average recoveries of USEPA Priority Pollutants from 5 sets of primary sewage sludge samples varied from -46 to 69%. Dewalle et al (9) reported recoveries of phenolics from sewage and sewage sludges ranging from almost zero to more than 200%. Canadian data indicates extremely poor recoveries of phenolics (less than 20%) from sludge, with recoveries of other toxic organics ranging from less than zero to 70% (7). Because of these problems, it is not possible to make definitive statements concerning the amounts and fate of toxic organics in sewage sludges. However, it appears that much of the data underestimate actual concentrations.

3. EXISTING DATA BASE

The main removal mechanisms for toxic organic compounds across waste-water treatment plants are volatilization, biological oxidation and accumulation in the sludge. Since many of the compounds are non-volatile and biorefractory, they would be expected to accumulate in the sludge and existing data confirms that this happens. Pesticides, herbicides and polychlorinated biphenyls (PCB's) were among the first organics of concern. Almost all sewage sludges have been shown to contain these compounds. More recently, concern has been expressed about other toxic organic compounds and researchers have attempted to determine whether they too accumulate in sludges.

The potential for toxic organics to accumulate in sludges can be assessed from a knowledge of the physical/chemical properties of the compounds (10). In general, non-polar biorefractory organics with low Henry's Law constants will accumulate. The literature indicates that concentration factors (sludge concentration divided by influent wastewater concentration) can vary from one to four orders of magnitude. For complete mix activated sludge plants, concentration factors for compounds that are totally adsorbed on sludge vary from 5 000 to 20 000 (10). Because of these high concentration factors, it is common to detect compounds in the sludge that are not detected in the influent (11,12,13).

One of the most comprehensive surveys of toxic organics in sewage sludge was a USEPA study of forty Publically Owned Treatment Works (12). The

survey indicated the presence of 76 USEPA Priority Pollutants, ranging in
concentration from 1 to 70 000 μg/L. Assuming a sludge solids concentration
of 4%, these values reflect dry weight concentrations ranging from 0.025 to
1 550 μg/g. The organic compounds detected most frequently were
bis(2-ethylhexyl)phthalate, ethylbenzene, benzene, methylene chloride, 1,2
trans-dichloroethylene, anthracene, phenanthrene and phenol. The Metropolitan Sanitary District of Greater Chicago has also conducted a major survey of
toxic compounds in sewage sludges (11). Results from this study showed major
accumulations of pesticides, PCB's, phenols, and cyanides. Only eight base-
neutral extractable compounds were identified, with napthalene at 600 μg/L
being the highest reported value. Data from recent Canadian surveys by the
Ontario Ministry of the Environment and Environment Canada, including 13
sewage treatment plants, were comparable to those from the USA studies.

Selected data from the USA and Canadian surveys are shown in Table I.
The USA data were converted from wet weight concentrations (μg/L) to dry
weight concentrations (μg/g) to facilitate comparison with the Canadian
data. The data indicate accumulation of PNA's, phthalates and several other
compounds in sludges. Morever, there appears to be a relationship between
the types of compounds in sludge and the industries discharging to the sewerage system. For example, USA Plant #1 and Canadian Plant #2 receive wastewaters from coking and foundry operations and the sludges contained significant quantities of polynuclear aromatics (PNA's). Too few data were available to explore the relationship between the proportion of industrial influent to the sewerage system and the quantities of toxic organic compounds accumulated in sludges.

4. EFFECT OF SLUDGE PROCESSING ON TOXICS FATE AND MOBILITY

Experimental evidence indicates that non-volatile biorefractory organics
accumulate in sludges. This accumulation could be regarded as acceptable
"treatment" if the organics were to remain immobile, and were not solubilized
during subsequent sludge processing operations or following ultimate
disposal. Unfortunately, very little is known about the fate and mobility of
toxic organics in sludge. Processes which might affect their fate and
mobility include digestion, conditioning, dewatering and solidification.

Digestion is probably the most commonly used sludge stabilization process. It operates at a high solids retention time and provides ample opportunity for biological oxidation, volatile solids reduction and solubilization
of toxic organics. Limited data indicate that volatile compounds, with the

TABLE I. SEWAGE SLUDGE TOXIC ORGANIC DATA (μg/g) dry wt.)

PARAMETER	USA DATA (12)*			CANADIAN DATA**		
	Plant #4 Digested	Plant #12 Raw	Plant #1 Raw	Plant #1 Digested	Plant #2 Raw	Plant #3 Raw
Plant Flow (10^3 m^3/d)	318	141	345	40	264	64
% Industrial Flow	7	50	30	40	45	40
PNA Compounds						
Acenaphthylene	0.6				42	
Anthracene	8	83	27.7		311	
Phenanthrene	8	83	27.7	1	599	
Chrysene	1.3	0	8.5		60	
Benzo-a-pyrene					34	
Fluoranthene	1.1	9		2	334	
Fluorene	0.8		5.5	2	115	
Indeno(1,2,3,c,d)pyrene					38	
Napthalene	9.7	18	3.4	2	230	
Pyrene	2.0	25	13.2	3	236	
Other Compounds						
Chlorobenzene	5.5	2.2				
Toluene	40	225				
Nitrosodiphenylamine				3	8	29
Di-n-butylphthalate	0.7	36		38	57	134
Diethylphthalate		22			2	9
Di-n-octylphthalate				6	8	
Bis(2-ethylhexyl)phthalate	184	959	39	215	68	11
Isophorone						70
Dibenzofuran					64	
Carbazole					85	

* USA Plants #12 and #1 were primary sludges; USA Plant #4 was a mixture of primary and waste activated sludge.
** Mixtures of primary and waste activated sludge analyzed at the WTC.

exception of toluene, are either degraded, stripped to the atmosphere or transferred to the supernatant during digestion, whereas PNA's and phthalate are conservative (10). No data were found concerning the effect of diges-tion on toxics mobility.

Similarly, very limited information is available regarding the effects of sludge conditioning, dewatering and solidification on the fate and mobi-lity of toxic organics in sludges (10). Additional research is required to document effects and to optimize these processes for toxic organics manage-ment.

Environment Canada is funding a major study to determine the effect of anaerobic digestion, heat treatment and digestion plus polymer conditioning on the fate and mobility of selected toxic organics in sludge. The project was initiated in December, 1981 and the first phase, which comprised analytical methods assessment and sludge selection, was completed March 31, 1982. Canadian Sludge #2 (Table I) from a large municipal sewage treatment plant was selected and contains very high levels of PNA's, phthalates, and other compounds. The toxic organics to be monitored include acenaphthylene, benzo-a-pyrene, fluoranthene, fluorene, pyrene, di-n-butylphthalate, carbozole and dibenzofuran. Raw sludge will be subjected to laboratory digestion and heat treatment, and each treatment will be run in triplicate on three sludge samples collected at four-month intervals in order to assess variations in toxic organic contents with time and treatment. Toxics mobility will be assessed using the proposed EPS Leaching Procedure (14). Data from this study will be available by March 31, 1983, and should allow engineering judgements to be made regarding the impact of sludge processing on utilization and disposal options.

5. AGRICULTURAL UTILIZATION

Land spreading of sewage sludge is practiced in many countries. To date, the major concerns with respect to this practice have been health risk from pathogenic organisms and transfer of heavy metals from the sludges to soils and crops. A large amount of data exists on the fate of pathogens and metals in sludges applied to land, and guidelines and regulations have been developed using these data. By contrast, very limited data is available on the fate of toxic organic compounds and these materials are largely ignored in the guidelines.

Guidelines for sludge utilization in Ontario and Alberta specify limitations for pathogens and metals. However, the Ontario Guidelines recognize the potential hazards associated with organics such as PCB's, PBB's and phthalates in sludge, and indicate that, as the data base is expanded, future guidelines will address these and other "potentially toxic and persistent materials". The USEPA has proposed limits for PCB's, pesticides and other persistent organics in addition to limits for metals and pathogens. The authors are not aware of any European guidelines that address addition of toxic organic compounds in sludge to land.

Several classes of organic compounds of concern relative to land application of sludge are pesticides, PCB's and other chlorinated aromatics, and PNA's. The behaviour and decomposition pathways of pesticide residues in soils have been studied for many years (15). Degradation can occur chemically and microbiologically and the rate is affected by a variety of factors including chemical structure and solubility, soil properties and temperature. Plants take up many pesticides through their roots and trans-locate them to the above ground vegetation. Considerable research has also been conducted on the fate of PCB's in soils (16). Studies incorporating PCB's in normal agricultural soils have indicated some degradation and sub-stantial volatilization over relatively long time periods. However, other studies indicate that adaptation of soil microorganisms can lead to rapid degradation of the less chlorinated compounds, thereby reducing volatiliza-tion. Carrots which exhibit a high propensity to take up lipophilic sub-stances, may contain PCB's in the ppm range; almost all of which is associ-ated with the roots and peel.

The fate of other chlorinated aromatics and PNA's in soils has received considerably less attention than pesticides and PCB's. There is evidence (16) that many microorganisms, isolated from soil or polluted waters, can degrade up to 3 ring-membered PNA's. Higher PNA's such as 5 ring-membered compounds are not known to be utilized by microbes as a sole carbon source, however, there is evidence that they can be cometabolized and cooxidized with other organics. No information is available on the effects of PNA's on plants or soil physical properties. Pentachlorophenol (PCP) and other halo-genated phenols have been shown to decompose in soil and the half-life for PCP at the 100 ppm level in soil at 30°C has been determined as approxi-mately 40 days. The half-lives of pentachloronitrobenzene (PCNB) and hexa-chlorobenzene (HCB) averaged 270 and 1 530 days, respectively, when both com-pounds were applied to soil at 10 kg/ha. The treated soil was maintained at 60% field capacity and 18-20°C in the laboratory. The calculated half-life of PCNB for 22 field samples that had received this compound for 2-10 years ranged from 117-1 059 days with an average of 434 days. Both PCNB and HCB may be taken up by plants and there is evidence that PCNB may be transformed to pentachloroaniline and pentachlorothioanisole in plants (16)

Much of the available information on the fate of toxic organics applied to land has been derived from laboratory culture experiments or reflects a one-time response of the plant-soil system to a single compound. Major research programs are required to determine the effects on soils and plants of repeated applications of sludges containing many compounds.

A research program has been initiated at the WTC to determine the fate of toxic organic compounds in Canadian municipal and industrial sludges ap-plied to land. A preliminary greenhouse study is in progress with three municipal sludges, an oil refinery biosludge, an oil refinery tank bottoms sludge, a petrochemical industry biosludge, and a coke plant biosludge. One of the municipal wastewater treatment plants receives a large proportion of its influent from a petrochemical industry and another from coking and foundry operations. Analyses indicate that all of the sludges contain organic compounds of concern, however, the municipal sludge with the least industrial input contains low concentrations of these compounds. This sludge has promoted excellent growth of several field and vegetable crops in previous lysimeter experiments and was included in the present study as a sludge control treatment. The study also includes a commercial fertilizer control treatment.

The sludges were mixed completely with Conestoga loam soil at rates of 20, 100 and 200 tonnes/ha. The mixtures were incubated moist in ceramic pot for three weeks and then Swiss chard was planted. Although germination occurred in all treatments, there has been poor growth or death of the seedlings in several of the high sludge application rate treatments. The cause of poor survival and growth has not been established. It might be due to toxic organic compounds in the sludges, or to other causes such as high concentrations of soluble salts and ammonia. Following harvest, analyses will be conducted to determine whether plant uptake and/or soil degradation of the toxic organic compounds has occurred. Upon completion of this preliminary study, further greenhouse experiments will be conducted with Canadian Sludge #2 (Table I).

6. SLUDGE INCINERATION

Incineration is widely used for the ultimate disposal of sewage sludges. In Ontario, 40% of all sewage sludge is incinerated (17). Although incinceration has been practiced extensively, recent problems with compounds such as PCB's, Mirex and Kepone have aroused concern about the fate of toxic organics. Existing data bases define the fate of heavy metals in sludges during incineration (18,19), however, little information exists for toxic organics. One of the few classes of compounds for which data does exist is PCB's. Destruction of PCB's in cement kilns has been shown to be an economically viable and environmentally sound control strategy (20).

The need for additional toxics "incinerability" data has been recognized and studies are in progress. The USEPA contracted the University of Dayton Research Institute to develop laboratory facilities to determine the fate of specific toxic organic compounds during incineration. The result has been development of the Thermal Decomposition Analytical System and the Thermal Destruction Unit (21). These units have been used to generate thermal decomposition and destruction profiles for compounds such as PCB's, Kepone, Mirex and DDT (22,23).

An informal cooperative agreement between the USEPA and Environment Canada exists to liaise in the generation of laboratory thermal destruction profiles. A laboratory-scale system, similar to the Thermal Destruction Unit, has been assembled at the WTC. The system, shown schematically in Figure I, comprises a vapourization chamber (actually a Thermogravimetric Analyser), a high temperature reactor and a gas collection and analysis unit. Up to 200 mg of sample is inserted into the vapourization chamber where the temperature can be controlled to give the required volatilization rate. A carrier gas transports the vapours to the reactor, which consists of a folded quartz tube, 300 mm long by 2 mm diameter. Here, under plug flow conditions, gas residence times of 0 to 5 seconds can be maintained. The products of combustion are either analysed on-line using a gas chromatograph or are trapped for subsequent analysis.

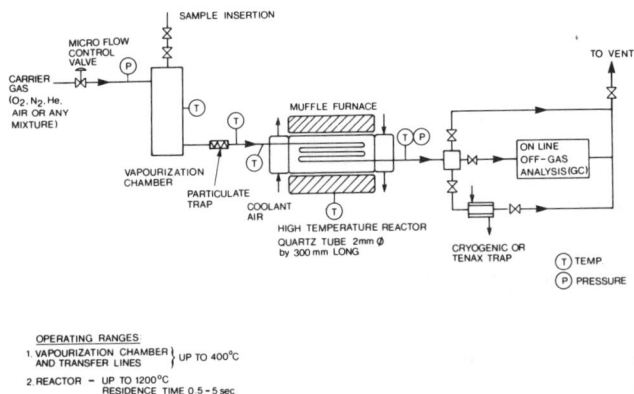

Figure I.   WTC THERMAL DATA GENERATION SYSTEM.

The equipment at the WTC is currently being used to establish the fate of penta-, tetra- and trichlorophenols when wood preserving industry wastewater treatment plant sludges are incinerated. The potential formation of dibenzo-p-dioxins during incineration of these compounds is a major concern. Additional research is planned to establish the fate of PNA's, pesticides, and other compounds present in municipal sludges. The major aim of the studies is to determine whether sludge incineration can provide safe, economic control of toxic organics.

7. CONCLUDING REMARKS

There is conclusive evidence that a wide variety of toxic organic compounds can occur in Canadian and USA sewage sludges. Limited information is available concerning the nature and properties of these compounds, however, much more information is required to define their fate during processing, land utilization and incineration of sludges.

Research in progress at the WTC should provide much needed information on the fate of toxic organics in sewage sludge and should facilitate decisions concerning processing, utilization and disposal options to minimize environmental damage. Several possible options include; pretreatment of industrial effluents to remove toxics prior to entry into the domestic sewer system, sludge processing to decompose toxics, and incineration to thermally destroy toxics.

REFERENCES

1. Chen, F., Private communication (1982).

2. Federal Register, Guidelines Establishing Test Procedures for the Analysis of Pollutants; Proposed Regulations – Part III 40 Code of Federal Regulations 136, Vol. 44(233): 69464 (Dec. 3, 1979).

3. Dacre, J.C., Potential Health Hazards of Toxic Organic Residues in Sludge. In Sludge–Health Risks of Land Application, Eds., G. Bitton, B.L. Damron, G.T. Edds and J.M. Davidson, Ann Arbor Science Publishers Inc., pp. 85-102 (1980).

4. Monteith, H.D. and J.P. Stephenson, Development of an Efficient Sampling Strategy to Characterize Digested Sludges. COA Research Report No. 71, Ottawa (1978).

5. USEPA, Analytical Procedures for Determination of Organic Priority Pollutants in Municipal Sludges. EPA-600/2-80-030 (1980).

6. Midwest Research Institute, Development of Analytical Test Procedures for the Measurement of Organic Priority Pollutants in Sludges and Sediments. Special Report #4, EPA Contract No. 68-03-2695 (1980).

7. Conn, K., Private Communication (1982).

8. Petrasek, A.C., B.M. Austin, T.A. Pressley, L.A. Winslow and R.H. Wise, Behavior of Selected Organic Priority Pollutants in Wastewater Collection and Treatment Systems. Presented at 53rd Annual WPCF Conference, Las Vegas, Nevada (1980).

9. Dewalle, F.B., D.A. Kalman, R. Dills, D. Norman, E.S.K. Chian, M. Giabbai and M. Ghosal, Presence of Phenolic Compounds in Sewage Effluent and Sludge from Municipal Sewage Treatment Plants. Water Sci. Tech., 14:143-150 (1982).

10. Bridle, T.R., The Impact of Toxic Organics on Residue Management. Unpublished report, WTC, Burlington, Ontario (1981).

11. Lue-Hing, C., R.I. Pietz, J.R. Peterson, G.F. Richardson and D.T. Lordi, Effects of Priority Pollutants on the Disposal of Sludges from Publicly owned Treatment Works. The Metropolitan Sanitary District of Greater Chicago, Dept. of Research and Development, Report No. 81-4 (1981).

12. USEPA, Fate of Priority Pollutants in Publicly Owned Treatment Works. Interim Report EPA-440/1-80-301 (1980).

13. Bridle, T.R., W.K. Bedford and B.E. Jank, Biological Treatment of Coke Plant Wastewaters for Control of Nitrogen and Trace Organics. Presented at 53rd Annual WPCF Conference, Las Vegas, Nevada (1980).

14. Environment Canada, A Proposed Procedure for the Development of a Canadian Data Base on Waste Leachability. Internal document, WTC, Burlington, Ontario (1981).

15. Leonard, R.A., G.W. Bailey and R.R. Swank, Jr., Transport, Detoxification, Fate and Effects of Pesticides in Soil and Water Environments. In Land Application of Waste Materials, Soil Conservation Soc. America, Ankeny Ia., pp. 48-78 (1976).

16. Overcash, M.R. and D. Pal, Specific Organic Compounds. In Design of Land Treatment Systems for Industrial Wastes - Theory and Practice, Ann Arbor Science Publishers Inc. pp. 221-334 (1979).

17. Black, S.A. and N.W. Schmidtke, Practices and Trends in Sewage Sludge Utilization and Disposal. In Proceedings of the First Canada/Germany Wastewater Treatment Technology Exchange Workshop held at the WTC, Burlington, Ontario, pp. 76-99 (1979).

18. Campbell, H.W., P.J. Crescuolo and T.R. Bridle, Fate of Heavy Metals and Potential for Clinker Formation During Pilot Scale Incineration of Municipal Sludge. Wat. Sci. Tech., 14:463-473 (1982).

19. Dewling, R.T., R.M. Manganelli and G.T. Baer, Fate and Behavior of Selected Heavy Metals in Incinerated Sludge. JWPCF, 52(10):2552-2557 (1980).

20. MacDonald, L.P., D.J. Skinner, F.J. Hopton and G.H. Thomas, Burning Waste Chlorinated Hydrocarbons in a Cement Kiln, EPS 4-WP-77-2 (1977).

21. USEPA, Design Considerations for a Thermal Decomposition Analytical System. EPA-600/2-80-098 (1980).

22. USEPA, Laboratory Evaluation of High Temperature Destruction of Kepone and Related Pesticides. EPA-600/2-76-299 (1976).

23. USEPA, Laboratory Evaluation of High Temperature Destruction of PCB's and Related Compounds. EPA-600/2-77-228 (1977).

TOXIC ORGANIC COMPOUNDS IN TOWN WASTE MATERIALS: THEIR ORIGIN, CONCENTRATION AND TURNOVER IN WASTE COMPOSTS, SOILS AND PLANTS

H. HARMS and D. R. SAUERBECK

Institute of Plant Nutrition and Soil Science
Federal Research Center of Agriculture Braunschweig-Völkenrode (FAL)
Federal Republic of Germany

Summary

Beside heavy metals, a number of xenobiotic organic substances like polychlorinated biphenyls and polycyclic aromatic hydrocarbons are formed or increased as a result of modern civilization. Since municipal wastes like town garbage composts and sewage sludge are used as soil amendments in agriculture, the potential incorporation of xenobiotic organics into the human food chain must be observed. Fortunately, in contrast to the more sensitive aquatic ecosystem, most agricultural investigations showed only a limited significance of these compounds in soils and in food. There were, however, exceptions which are of considerable public and administrative concern.

Since town waste materials will never be free from undesirable contaminants, their use in agriculture needs to be monitored and their composition to be controlled. A regular surveying of their quality will also result in a better identification and elimination of hitherto unknown sources of critical compounds. In agricultural practice, however, it will be inevitable to compromise, trying to keep xenobiotics away from soils as far as possible, but also not to exaggerate their significance in foods. Crops like small grains are known to have low absorption of such organics and not to translocate them into their edible parts. If town wastes are recycled in areas producing mainly such crops, the risks involved are considered to be relatively low.

1. INTRODUCTION

Considerable quantities and a wide spectrum of possibly toxic organic substances are liable to be released into the environment as a result of either industrial or agricultural activities. It is essential, therefore, to understand the fate and the ecological effects of such compounds in the different environmental compartments. As far as the agro-ecosystems are concerned, the use of pesticides and the intentional application of town waste materials are often considered to be the main sources of contamination.

Such town waste materials are introduced into field soils as either garbage composts or sewage sludge or composted mixtures of both. According to our present knowledge, out of a whole range of organic town waste con- taminants the polychlorinated_biphenyls (PCBs) and the polycyclic aromatic_hydrocarbons (PAHs) deserve the biggest attention, either because of their quantities present or of their stability, and last but not least due to their possible cancerogenic effects. This present review will, therefore, be limited to these two main groups, trying to describe their sources and transfer as well as their turnover behaviour in soils and in plants.

2. SOURCES AND TRANSFER TO THE ENVIRONMENT

The polychlorinated_biphenyls (PCBs) comprise a multitude of biphenyl homologues differing in their degree of chlorine substitution. Their commercial production started about 50 years ago. Since then they were increasingly used for a whole range of industrial and electro-technical purposes. Unfortunately, their most critical importance as an environmental contaminant has not been recognized before 1966, when JENSEN and others discovered these compounds in fish and in fish-eating birds (14, 28, 42, 60). High contents were found not only in industrial wastes, but also in river sediments and in packing materials, indicating an almost ubiquitary oc- currence with particularly large concentrations in urban and industrial areas (1, 43, 44).

Due to their utilization in various consumption goods, such as e. g. paints, plastics and self-copying papers, the PCBs also enter the municipal waste waters and solid wastes, from where they are eventually transferred to other environmental compartments. Similar to the well-known DDT phenomenon, they are strongly enriched particularly in the aquatic food chains, so that the organisms at the end of such chains may contain considerable PCB-concen- trations, although the original contents in the primary media soil, air and water were comparably low. This of course caused a lot of concern about food wholesomeness and human health, and a number of US rivers and lakes had in fact to be closed to commercial fishery (1).

The environmental contamination with polycyclic aromatic hydro-carbons (PAHs) can be traced back mainly to industrial, household and motor vehicle exhausts or related activities of modern civilization (13, 16). The pyrolysis and/or incomplete charring of organic substances at temperatures between 500 and 700 °C release an array of more than 160 PAHs of most different composition (59). The same phenomenon also occurs during tobacco smoking (15). As a consequence of their adsorption to ash particles, smoke and soot, these PAHs are subsequently transferred to the environment and finally deposited somewhere else (50).

MÜLLER et al. investigated the PAH content of sediment cores from the Lake Constance and discovered that its manifold enrichment in the last 75 years corresponded with the increasing combustion of fossil fuel (especial-ly coal) during this time (37). The distribution of PAHs occurs mainly by waste water and air, the rain washes them down from there into waters and soils (32). As with the heavy metals and PCBs, the PAH contamination is highest in urban and industrial zones. They are regularly found in both plants and in animal tissues, and can be particularly accumulated from waste waters through the aquatic food chains by fish (45).

3. CONTENTS IN SOILS AND IN TOWN WASTE MATERIALS

The translocation of polychlorinated biphenyls by wind, rain and water eventually leads to the contamination of soils with such com-pounds as well. The higher the extent of chlorination, the longer these compounds persist. 95 % of Arochlor 1254 (a commercial mixture of tetra-, penta- and hexachloro-biphenyl isomers) has been recovered from two soils rich in organic matter by IWATA et al. one year after its addition. In a loamy-sand soil, however, the tetra- and pentachloro-isomers decreased considerably, whereas the more chlorinated biphenyls remained (24). Ac-cordingly, the breakdown of PCBs depends on the type of soil, the reaction time, the environmental conditions, and the extent of chlorination. How-ever, almost no precise figures are known to the authors regarding the actual PCB contents of agricultural soils.

More data exist of the PCB concentrations in waters and waste water sludges. The contents of clear waters are usually low, because the PCBs are adsorbed by the particulate material suspended or settling there. Conse-quently, the PCB concentrations in sewage sludges exceed those of waste waters many times. According to JELINEK (27) US sludges were found to con-tain up to 352 ppm PCBs in the dry matter. OTTE and LA CONDE (41) studied nine landspreading sites representing a wide range of sludge application rates and detected in 74 % of the sludge samples significant quantities (5100 ppb) of PCBs. 44 % of the surface soil samples contained levels of 300 ppb (41). 1 - 185 ppm were reported from Scotland by HOLDEN (23) and up to 26 ppm by JANSSON (26) from Sweden. Considerably less has been found in Dutch sludges ranging from 0.2 - 10 ppm, with a median of 0.6 and an average content of about 1 ppm (19). Incidentally, the contents of chlorinated pesticides were usually shown to be at least one order of magnitude less (27).

German municipal wastes were analyzed by MÜLLER et al. (39), showing PCB contents of about 2 ppm in sludge and between 0.4 and 9.7 ppm in the garbages. Town garbage and sewage sludge-garbage composts contained between 2 and 5 ppm, while the potting media and garden moulds formed from them showed 0.4 - 2.5 ppm (Table I).

Table I: PCB-content (ppm) in municipal wastes, sewage sludges and municipal waste composts (MÜLLER et al., 1974)

sample	range min.	range max.	number of samples
town waste	0.4	9.7	10
town waste/ sewage sludge	2.1	2.5	2
sewage sludge	1.8	2.0	2
fresh compost		1.9	1
ripe compost	2.9	5.0	4
garden mould	0.4	2.5	3

The normal occurrence of polycyclic aromatic hydrocarbons both in countryside and agricultural soils has been shown by BLUMER (3) as early as 1961. He found between 40 and 1300 ppb benzo(a)pyrene in forest soils and about 90 ppb in arable fields. German and European soils contain PAHs in the same order of magnitude, although the range varies from zero to a maximum of 5000 ppb (29, see Table II).

The PAHs in soils represent a complex assemblage of homologous compounds. BLUMER et al. (4) conclude from their findings that such potentially carcinogenic and mutagenic hydrocarbons occurred on the earth's surface already since geological time spans, because they are always formed during natural fires. Most recent data of MATZNER et al. (34) show a considerable accumulation of 4 different PAHs in the humus layer of forests as compared with the underlying mineral soil. This is attributed to the great filtering action of trees upon particulates and aerosols, which results in higher deposition rates under spruce than in beech forest stands.

Table II: Benzo(a)pyrene and polycyclic aromatic hydrocarbons (sum of 6 compounds) in soils (KUNTE, 1977)

A. Benzo(a)pyrene range μg/kg	number of samples	B. sum of 6 PAHs range μg/kg	number of samples
< 2	0	< 50	0
2 - 10	14	50 - 100	6
10 - 30	19	100 - 300	23
30 - 50	4	300 - 500	8
50 - 100	5	500 - 1000	5
100 - 300	3	1000 - 3000	4
300 - 500	4	3000 - 5000	3
500 - 1000	0	> 5000	1
> 1000	1		

The relatively large concentration of PAHs found in town waste products has raised the suspicion that their agricultural utilization might perhaps influence food quality and human health (57). A number of investigations have therefore been done in order to measure the possible accumulation of PAHs from town composts and sludges in soils and plants (6, 8, 9, 30, 48). ELLWARDT for instance from our own institute analyzed a soil which received 100 t sludge or town compost dry matter but surprisingly discovered about 160 different PAHs even in the untreated control (8). Most of these PAHs, however, are still unidentified compounds. A quantitative measurement of the eight most interesting PAHs in sewage sludge and in fresh and decomposed waste composts (8) revealed overall contents between 7 and 8 ppm (Table III).

Table III: Amounts (in ppb) of various polycyclic aromatic hydrocarbons in sewage sludge and waste composts from Bad Kreuznach (ELLWARDT, 1977)

Polyarene	sewage sludge	fresh waste compost	decomposed waste compost
Benzo(b,k,j)fluoranthene	2962	3400	3270
Benzo(e)pyrene	1247	1750	1520
Benzo(a)pyrene	806	680	655
Perylene	250	131	109
Indeno(1,2,3-cd)pyrene	281	457	311
Dibenz(a,h)anthracene	684	785	847
Dibenz(a,c)anthracene	257	134	133
Benzo(ghi)perylene	610	802	823
sum	7097	8134	7668

The compound benzo-fluoranthene turned out to be the most predominant one, while benzo(a)pyrene and dibenz(a,h)anthracene contributed about 10 % each. Other investigators found similar, several of them even higher contents (6, 33), but with a few exceptions the orders of magnitude corresponded reasonably well.

There is little doubt that the application of such town waste materials raises the PAH contents of the treated soils. As the experiments at our own institute indicate (Table IV, ELLWARDT, 1977), the amounts of seven out of the 8 PAHs mentioned above increased up to twofold compared with the control. The data of many other investigators point to the same direction, although the variations were large depending on the local and experimental conditions (6, 9, 46, 58). Some of these variations have been explained by a de-novo-formation of PAHs during the microbial breakdown of organic matter (5), but there were also some opposite findings of microbial PAH degradation (55, 58).

Table IV: Amounts (ppb) of various polycyclic aromatic hydrocarbons in soils treated with sewage sludge and waste composts (ELLWARDT, 1977)

Polyarene	test plot	plot treated with sewage sludge	plot treated with fresh compost	plot treated with decom- posed compost
Benzo(b,k,j)- fluoranthene	112	129	144	176
Benzo(e)pyrene	48.3	52.4	78.0	88.0
Benzo(a)pyrene	23.2	31.5	89.2	40.7
Perylene	4.2	5.9	7.2	7.0
Dibenz(a,h)- anthracene	17.5	17.1	28.3	29.6
Dibenz(a,c)- anthracene	38.1	43.6	57.5	63.4
Benzo(ghi)- perylene	13.4	9.3	24.4	27.5

This can be concluded for instance from investigations of MARTENS (33) at our research centre, according to which no accumulation of any of six 4-6 ring PAHs was found during town-waste compost fermentation. As there was a considerable loss of organic matter by mineralization, the more persistent PAHs should have become relatively enriched during this process. The fact that this was not the case indicates that part of these PAHs had been degraded.

A direct proof of the ability of fermenting composts to degrade PAHs has been given by the same author (33) with ^{14}C-labelled PAHs. An at the first glance amazing result was that the ripe composts showed much higher rates of $^{14}CO_2$-release than the corresponding assays with the fresh material did, although the microbial activity here was much higher than there. This breakdown was the more pronounced, the less aromatic rings the labelled compounds contained. This corroborated the analytical findings according to which the relation of e. g. benz(a)anthracene versus the more complex 5-6 ring PAHs was lower in the ripe composts.

The low capacity of fresh composts to degrade most xenobiotics has already been pointed out earlier by MOLLER et al. (39). This may, however, be due to the use of fresh wastes in their model composting unit. Hence, their conclusion that the fermentation would not reduce the xenobiotics in town waste should not be generalized because the environmental conditions for and the composition of the microbial populations vary with the extent of decomposition (39). The breakdown of PAHs added to soils during the experiments of LÜW (31), however, was rather low and appeared to occur primarily by the co-metabolism of a PAH-adapted microflora. This result contrasts with earlier findings of SHABAD (45) and SIEGEL (46) who found a very quick reduction of the benzo(a)pyrene content in soils within a few days.

Another important result of MARTENS's (33) investigations was that in spite of considerable variations in the total PAH contents in the different composts, the relative ratios between the individual PAHs remained rather constant. This may permit the selection of one or a few individual PAHs like benzo(a)pyrene to represent the whole group instead of analyzing a large number of most complex compounds.

4. UPTAKE AND METABOLISM BY PLANTS

The ubiquitary distribution of the xenobiotics mentioned and their proven or assumed toxicities, together with their occurrence in soils and even more so their possible accumulations from town waste materials make it necessary to know, whether plants do absorb or perhaps even accumulate such contaminants which are then passed on to animals and to man. As some of these compounds are liable to further changes into even more toxic metabolic products (20), an important question is also if such undesirable transformations occur within living plants.

An evaluation of the limited knowledge available then about waste-water derived polychlorinated biphenyls has been given by RHODE (43) in 1975. As mentioned earlier, the PCBs like other lipophilic substances are strongly absorbed by the algae and other aquatic plants. Their accumulation factors from ocean water were shown to be about 1000fold, with the result that cell growth and division as well as their chlorophyll and RNA-synthesis could be strongly impeded (36). From there a large number of spine-less small animals become contaminated several 10,000fold as compared with the water in which they live. The passage of this contamination through predatory food chains has already been mentioned before (1).

The uptake of PCBs by higher terrestric plants was first studied by IWATA et al. (25) in 1974. They planted carrots on soil which was artificially enriched with 100 ppm Arochlor 1254. The chromatographic analysis revealed a preferential uptake of isomers of low chlorination, i. e. the same compounds which gradually disappeared from the soil during its incubation. 97 % of this ad- or absorbed PCB remained in the outer bark of the roots and did not penetrate further on. The isomer concentrations found in these outer root layers did not exceed 30 - 50 % of those in the surrounding soil.

It should be understood, however, that this experimental soil contained 100 ppm PCBs, which is 20 - 50 times more than what is found in town-waste derived composts (43). 50 tons of such composts per hectare would enrich the soil with only 0.064 ppm. Hence, the information gained from this model experiment is of theoretical significance only. NOREN (40) found only a few samples out of 2500 plant food items in Sweden which did in fact contain measurable amounts of PCBs. The situation should nevertheless be monitored, keeping in mind that at least water plants are able to dechlorinate some PCBs and to transform them into hydroxy-derivatives (35).

Green forage and roughages from sludge-treated land may also lead to a direct ingestion of sludge-derived PCBs by animals, which are also known to ingest considerable amounts of soil as such (7). The contents of PCBs in the milk fat of cattle were in fact shown to respond to such feed contaminations (27), so that the US-FDA recommends a maximum permissible content of not more than 10 ppm PCBs in agriculturally usable sludges (7).

Apart from their heavy metal content, the most serious objections against the agricultural utilization of municipal waste composts and sludges refer to their unobjectionable contents of polycyclic aromatic hydrocarbons (57). Although these contents are usually in the ppb-range only, the cancerogeneity of some of these compounds still justifies scientific concern and public care.

Contrary to the PCBs, a rather considerable number of investigations have already dealt with the contents and the metabolism of PAHs in plants. As early as 1966 GRÄF and DIEHL (10) published results which proved the existence of several PAHs in various plants. Their original assumption was that these aromatic compounds are naturally synthesized in plants and might even act as growth hormones (11). The biosynthetic formation of PAHs was in

fact shown by BORNEFF et al. (5) in lower plants. This led to the conclusion
that the PAHs found in mineral oil would be of biological origin too (12).

WAGNER et al. (55, 56) however do not believe that higher plants can
form benzo(a)pyrene or benzo(b)fluoranthene metabolically themselves, and
GRIMMER and DOWEL (17) in fact did not find any PAHs in lettuce, rye, soybeans and tobacco grown in a most carefully filtered growth chamber air. The
control plants from fields and ordinary greenhouses, however, contained PAHs
in the usual low concentrations, so that their uptake could be assumed to
take place from the air.

WAGNER et al. (54, 55, 56) were the first ones in Germany who studied
the uptake of PAHs added to soils. They observed an uptake by wheat, rye,
maize, barley and carrots and even an impeded development of both roots and
shoots at high PAH concentrations. Their results were, however, not very
consistent and partly even contradictory, because in some cases the control
plants contained more than the ones which were supplied with 1 or in some
cases even with 10 mg PAHs per pot. The contents of their plants were at any
rate very low (usually less than 1 µg/100 g dry matter) and varied depending
on growth stages and the plant parts. The relatively far-going and alarming
conclusions of WAGNER (57) have therefore been debated a lot and could not
be supported by other German investigators. BORNEFF et al. (6), SIEGFRIED
(49), MÖLLER (38), SIEGEL (47, 48), ELLWARDT (8), and LINNE and MARTENS (30)
invariably agreed that the PAH contents of the above-ground plant parts
remain very low even at high concentrations in soils. The lowest contents
were consistently shown in the cereal grains, unless heavy aerial contaminations occurred (6).

More has been found in potato tubers and in radish or carrot roots, but
most of the PAHs remained in the outer bark which is peeled or scraped off
before human consumption (8, 38). A convincing explanation for this low uptake is the fact that the PAHs added to soils are most strongly fixed there
by clay minerals and organic matter (34, 46). The only exception known are
pure silica sand media without any humus contents, but even there the uptake by carrot plants was rather small and the translocation of benzo(a)-
pyrene into the leaves extremely slow (38).

All the PAH measurements made in plants from soils treated with town
waste have so far confirmed this conclusion. The results of ELLWARDT (8) in
Table V stand for a multitude of similar data from Germany, all of them
showing a most limited uptake of town-waste derived PAHs from the soil.
There is good reason for the belief that at least for large-surfaced leafy
vegetables the contamination by the aerial deposition of PAHs exceed their
uptake even from measurably town-waste contaminated soils. BLUM and SWARBRICK

Table V: Amounts (in ppb) of various polycyclic hydrocarbons in potatoes, oats and rye (ELLWARDT, 1977)

plant/ tissue	test plot untreated	plot with fresh compost	plot with decomp. compost
potatoes			
tubers	19.3	26.7	31.2
stem	104.4	110.7	125.4
oats			
grain	48.1	95.5	51.0
straw	32.8	81.6	78.3
rye			
grain	25.8	24.7	32.9
straw	27.1	32.0	28.0

(2) even denied any translocation of ^{14}C-labelled benzo(a)pyrene by three leafy plants from a nutrient solution and believe only in its irreversible deposition into the plant cuticle.

A quantitative determination of all PAHs existing is practically impossible. Many investigations were, therefore, restricted to the most representative or more toxic compounds like benzo(a)pyrene. Tracer experiments have confirmed that this substance is taken up by the roots and translocated in various directions (18). Shoot-applied benzo(a)pyrene can even move basipetally and disappear from the plants (18). According to HARMS (20, 22) the uptake rate of PAHs depends on their molecular size. More highly condensated ring systems like benzo(a)pyrene and dibenz(a,h)anthracene are very little absorbed and translocated by intact plants. The lower ring numbers of e. g. benzanthracene and anthracene result in considerably higher contents in the plant roots and shoots as indicated by their radioactivity.

These tracer studies also allowed the metabolism and turnover of such PAHs within plants to be followed. Experiments with sterile plants and with plant cell suspension cultures revealed that benzo(a)pyrene can be metabolized into oxygenated derivatives (21, 51). Like in plant microsomes (52), benzo(a)pyrene 3,6- and 1,6-quinone could be identified especially in the tissue of Chenopodium rubrum plants. These compounds are known to be even more potent carcinogens than the original benzo(a)pyrene. However, the work of v. TRENCK and SANDERMANN (53) indicates that such quinones are polymerized into the insoluble plant lignin fraction, which may be an important mechanism for its detoxification.

5. CONCLUSIONS
Beside heavy metals a number of xenobiotic organic compounds like PCBs and PAHs are formed or increased as a result of modern civilization. By the

release of industrial and municipal wastes they find their way into aquatic and terrestric ecosystems. Since municipal wastes like town waste composts and sewage sludge are recommended and used as soil amendments in agriculture, the potential incorporation of such xenobiotic organics into the human food chain must be observed. Fortunately, in contrast to the more critical aquatic ecosystem, most agricultural investigations revealed only a limited signifi-cance of these compounds in soils and in food production. There were, how-ever, exceptions which are of considerable public and administrative concern.

Since town waste materials will never be free from more or less undesi-rable organic contaminants, their use in agriculture needs to be monitored and their composition to be controlled. The limitation of PCB contents to a maximum value of 10 ppm in the US is a step towards this direction. A regular surveying of the quality of town waste composts and sewage sludges will also result in a better identification and elimination of hitherto unknown sources of critical compounds. There is no question that this is a much more satisfactory solution of the problem than to accept any risk of a gradual soil and/or plant contamination.

In practice it will always be necessary to compromise, trying to keep xenobiotics away from our soils as far as possible, but also not to ex-aggerate their significance in foods. As a measure of safety town refuse and sludge should not be recommended as amendments on soils used for vegetable crops. On the other hand, there are crops like small grains known to have a particularly low adsorption of such organics or not to translocate them into their edible parts. If town wastes are preferentially recycled in areas producing mainly such crops, the risks involved are considered to be relatively low.

References

(1) AHMED, K.: PCBs in the Environment. Environment 18, 6 - 11 (1976).

(2) BLUM, S. C.; SWARBRICK, R. E.: Hydroponic growth of crops in solutions saturated with [^{14}C-]Benzo(a)pyrene. J. Agric. Food Chem. 25, 1093 - 1096 (1977).

(3) BLUMER, M.: Benzpyrenes in soil. Science 134, 474 - 475 (1961).

(4) BLUMER, M.; YOUNGBLOOD, W. W.: Polycyclic hydrocarbons in soils and recent sediments. Science 188, 53 - 55 (1975).

(5) BORNEFF, J.; SELENKA, F.; KUNTE, H.; MAXIMOS, A.: Experimental studies on the formation of polycyclic aromatic hydrocarbons in plants. Environm. Res. 2, 22 (1968).

(6) BORNEFF, J.; FARKASDI, G.; GLATHE, H.; KUNTE, H.: Über die Beziehungen zwischen Müllkompostdüngung und krebserregenden Stoffen im Boden und in Nahrungspflanzen. Gießener Berichte zum Umweltschutz, Heft 2, 1 - 40 (1972).

(7) BRAUDE, G. L.; JELINEK, C. F.; CORNELIUSSEN, P.: FDA's Overview of the potential health hazards associated with the land application of municipal wastewater sludges. Proc. 1975 Natl. Conf. Municipal Sludge Management and Disposal. Information Transfer, Inc. Rockville, Maryland, pp. 214 - 217 (1975).

(8) ELLWARDT, P.-Chr.: Variation in content of polycyclic aromatic hydrocarbons in soil and plants by using municipal waste composts in agriculture. IAEA-SM-211/31 "Soil Organic Matter Studies", 291 - 297 (1977).

(9) FARKASDI, G.: Die Problematik der organischen Schadstoffe bei der Kompostierung von Siedlungsabfällen. Gießener Berichte zum Umweltschutz, Heft 4, 65 - 72 (1974).

(10) GRÄF, W.; DIEHL, H.: Über den naturbedingten Normalpegel cancerogener polycyclischer Aromate und seine Ursache. Arch. Hyg. 150, 49 - 59 (1966).

(11) GRÄF, W.; NOWAK, W.: Wachstumsförderung bei niederen und höheren Pflanzen durch cancerogene polycyclische Aromate. Arch. Hyg. 150, 513 - 528 (1967).

(12) GRÄF, W.; WINTER, Ch.: Benzpyren im Erdöl. Arch. Hyg. 152, 289 - 293 (1968).

(13) GRÄF, W.: Kanzerogene Substanzen in der Umwelt. Beiträge zum Umweltschutz II. Erlanger Forschungen, Reihe B. Naturwissenschaften 7, 19 - 33 (1975).

(14) GREICHUS, J. A.; GREICHUS, A.; EMERICK, R. J.: Insecticides, polychlorinated biphenyls, and mercury in wild cormorans, pelicans, their eggs, food and environment. Bull. Envir. Contam. Toxicol. 9, 321 (1973).

(15) GRIMMER, G.; GLASER, A.; WILHELM, G.: Die Bildung von Benzo(a)pyren und Benzo(e)pyren beim Erhitzen von Tabak in Abhängigkeit von Temperatur und Strömungsgeschwindigkeit in Luft- und Stickstoffatmosphäre. Beitr. Tabakforschg. 3, 415 (1966).

(16) GRIMMER, G.; HILDEBRANDT, A.; BÖHNKE, H.: Untersuchungen über die Belastung des Menschen durch Luftverunreinigungen. III. Zbl. Bakt. Hyg. 1 Abt. Orig. B. 158, 35 (1973).

(17) GRIMMER, G.; DÜVEL, D.: Cancerogene Kohlenwasserstoffe in der Umgebung des Menschen. Untersuchungen zur endogenen Bildung von polycyclischen Kohlenwasserstoffen in höheren Pflanzen. Z. Naturforsch. 25 b, 1171 - 1175 (1970).

(18) GROSSE, B.: Zur Analytik von 3,4 Benzpyren und dessen Aufnahme über Wurzel und Sproß höherer Pflanzen. Diss. TU München 1978.

(19) DE HAAN, S.: Gehalten van pesticiden en PCB in zuiveringsslib. Landbouwkundig tijdschrift 88, 21 - 27 (1976).

(20) HARMS, H.: Metabolisierung von Benzo(a)pyren in pflanzlichen Zellsuspensionskulturen und Weizenkeimpflanzen. Landbauforsch. Völkenrode 25, 83 - 90 (1975).

(21) HARMS, H.; DEHNEN, W.; MÜNCH, W.: Benzo(a)pyrene metabolites formed by plant cells. Z. Naturforsch. 32, 321 - 326 (1977).

(22) HARMS, H.: Aufnahme und Metabolismus polycyclischer aromatischer Kohlenwasserstoffe (PCKs) in aseptisch kultivierten Nahrungspflanzen und Zellsuspensionskulturen. Landbauforsch. Völkenrode 31, 1 - 6 (1981).

(23) HOLDEN, A. V.: Source of polychlorinated biphenyl contamination in the marine environment. Nature 228, 1220 - 1221 (1970).

(24) IWATA, Y.; WESTLAKE, W. E.; GUNTHER, F. A.: Varying persistence of polychlorinated biphenyls in Californian soils under laboratory conditions.

Bulletin of Environm. Cont. and Tox. 9, 204 - 211 (1973).

(25) IWATA, Y.; GUNTHER, F. A.; WESTLAKE, W. E.: Uptake of a PCB (Aroclor 1254) from soil by carrots under field conditions. Bulletin Environm. Cont. and Tox. 11, 523 - 528 (1974).

(26) JANSSON, S. L.: Klärschlammverwertung in der Landwirtschaft Schwedens. Landwirtschaftl. Forschung SH 27/I, 61 - 66 (1972).

(27) JELINEK, Ch.; BRAUDE, G. L.: Management of sludge use on land. J. Food Protection 41, 476 - 480 (1977).

(28) JENSEN, S.: A new Chemical hazard. New Scientist 32, 612 (1966).

(29) KUNTE, H.: Polyzyklische, aromatische Kohlenwasserstoffe in landwirt-schaftlich genutzten Böden. Zbl. Bakt. Hyg., I. Abt. Orig. B. 164, 469 - 475 (1977).

(30) LINNE, C.; MARTENS, R.: Überprüfung des Kontaminationsrisikos durch polycyclische aromatische Kohlenwasserstoffe im Erntegut von Möhren und Pilzen bei der Anwendung von Müllkompost. Z. Pflanzenernähr. Bo-denkd. 141, 265 - 274 (1978).

(31) v. LÖW, E.: Vorkommen und mikrobieller Um- und Abbau von aromatischen Polyzyklen im Boden und in Siedlungsabfällen. Diss. Univ. Gießen 1978.

(32) LUNDE, G.; GETHER, J.; GJOS, N.; LANDE, M. B.: Organic micropollutants in precipitation in Norway. Atmospheric Environm. 11, 1007 - 1014 (1977).

(33) MARTENS, R.: Concentrations and microbial mineralization of 4-6 Ring Polycyclic aromatic hydrocarbons in composted municipal waste. Chemo-sphere 1982 (in press).

(34) MATZNER, E.; HÜBNER, D.; THOMAS, W.: Content and storage of polycyclic aromatic hydrocarbons in two forested ecosystems in northern Germany. Z. Pflanzenernähr. Bodenkd. 144, 289 - 288 (1981).

(35) MOZA, P.; WEISSGERBER, J.; KLEIN, W.; KORTE, F.: Metabolism of 2,2-dichlorobiphenyl in two plant-water-soil-systems. Bulletin Environm. Cont. and Tox. 12, 541 - 546 (1974).

(36) MOSSER, J. L.; FISHER, N. S.; TENG, T. C.; WURSTER, Ch. F.: Poly-chlorinated biphenyls toxicity to certain phytoplankters. Science 175, 191 - 192 (1972).

(37) MÜLLER, G.; GRIMMER, G.; BÖHNKE, H.: Sedimentary record of heavy metals and polycyclic aromatic hydrocarbons in lake Constance. Naturwissen-schaften 64, 427 - 431 (1977).

(38) MÜLLER, H.: Aufnahme von 3,4 Benzpyren durch Nahrungspflanzen aus künstlich angereicherten Substraten. Z. Pflanzenernähr. Bodenkd. 139, 685 - 695 (1976).

(39) MÜLLER, W.; ROHLEDER, H.; KLEIN, W.; KORTE, F.: Modellstudie zur Ab-fallbeseitigung: Verhalten repräsentativer xenobiotischer Substanzen bei der Müllkompostierung. GSF Bericht Ö 104 (1974).

(40) NOREN, K.: - Arch. Hyg. Rada Toksikol. 24, 395 - 404 (1973).

(41) OTTE, A. D.; LA CONDE, K. V.: Environmental assessment of municipal sludge utilization at nine locations in the United States. In: Food, Fertilizer and Agricultural Residues. Proceedings of the 1977 Cornell Agricultural Waste Management Conference (Loehr, R. C. - editor; p. 135 - 146 (1977). Ann Arbor Science Publishers Inc.

(42) PRESTT, I.; MOORE, N. W.; JEFFRIES, D. J.: Polychlorinated biphenyls in wild birds in Britain and their avian toxicity. Envir. Poll. 1, 3 (1970).

(43) ROHDE, G.: Polychlorierte Biphenyle (PCBs). Bedeutung, Verbreitung und Vorkommen in Müll- und Müllklärschlammkomposten sowie in den damit gedüngten Böden. ANS-Mitteilungen SH Nr. 1, 1 - 15 (1975).

(44) RISEBROUGH, R. W.; RIECHE, P.; PEACALL, D. P.; HERMAN, S. G.; KIRVEN, M. N: Polychlorinated biphenyls in the global ecosystem. Nature 220, 1098 - 1102 (1968).

(45) SHABAD, L. M.: Circulation of carcinogenic polycyclic aromatic hydrocarbons in the human environment and cancer prevention. JNCI 64, 405 - 410 (1980).

(46) SIEGEL, O.: Beurteilung der verschiedenen Kompostierungsverfahren und der Verwendung von Kompost bezüglich der Anreicherung von Schwermetallsalzen und cancerogenen Stoffen im Boden. Stuttgarter Berichte zur Abfallwirtschaft Bd. 6 (1974), E. Schmidt-Verlag Berlin.

(47) SIEGEL, O.: Über den Gehalt von polycyclischen aromatischen Kohlenwasserstoffen in Weinbergsböden, Rebblättern und Wein. Festschrift zum 100jährigen Bestehen der LUFA Speyer (1975 a), Pfälz. Verlagsanstalt Landau.

(48) SIEGEL, O.: Über die Aufnahme von polycyclischen aromatischen Kohlenwasserstoffen durch Kulturpflanzen. Festschrift zum 100jährigen Bestehen der LUFA Speyer (1975 b), Pfälz. Verlagsanstalt Landau.

(49) SIEGFRIED, R.: Einfluß von Müllkompost auf den 3,4 Benzpyren-Gehalt von Möhren und Kopfsalat. Naturwissenschaften 62, 300 (1975).

(50) STEUBING, L.: Natürliches und immissionsbedingtes Vorkommen von Benzo-(a)pyren. Angew. Botanik XLV, 1 - 10 (1971).

(51) v. d. TRENCK, K. T.; SANDERMANN, H.: Metabolism of benzo(a)pyrene in cell suspension cultures of parsley (Petroselinum hortense, Hoffm.) and soybean (Glycine max L.). Planta 141, 245 - 251 (1978).

(52) v. d. TRENCK, Th.; SANDERMANN, H.: Oxygenation of benzo(a)pyrene by plant microsomal fractions. FEBS Letters 119, 227 - 231 (1980).

(53) v. d. TRENCK, Th.; SANDERMANN, H.: Incorporation of benzo(a)pyrene quinones into lignin. FEBS Letters 125, 72 - 76 (1980).

(54) WAGNER, K. H., WAGNER-HERING, E.; BUCHHAUPT, K.: Üben 3,4 Benzpyren und 3,4 Benzfluoranthen einen wachstumsfördernden Effekt auf Pflanzen aus? Z. Pflanzenernähr. Bodenkd. 123, 186 - 196 (1969).

(55) WAGNER, K. H.; SIDDIQI, I.: Der Stoffwechsel von 3,4 Benzpyren und 3,4 Benzfluoranthen im Sommerweizen. Z. Pflanzenernähr. Bodenkd. 127, 211 - 218 (1970).

(56) WAGNER, K. H.; SIDDIQI, I.: Die Speicherung von 3,4 Benzfluoranthen im Sommerweizen und Sommerroggen. Z. Pflanzenernähr. Bodenkd. 130, 241 - 243 (1971).

(57) WAGNER, K. H.; SIDDIQI, J.: Gefährliche Stoffe in Bodenverbesserungsmitteln. Naturwissenschaften 60, 160 - 161 (1973).

(58) WAGNER, K. H.; VONDERHEIT, C.: Polycyclische aromatische Kohlenwasserstoffe in technisch hergestellten Komposten. Naturwissenschaften 65, 491 (1978).

(59) YOUNGBLOOD, W. W.; BLUMER, M.: Polycyclic aromatic hydrocarbons in the environment: homologous series in soils and recent marine sediments. Geochem. Cosmochem. Acta 39, 1303 - 1314 (1975).

(60) ZITKO, V.; CHOI, P. M. K.: PCB and other industrial halogenated hydrocarbons in the environment. Fish. Res. Bd. of Canada Technical Report 272, 1 (1971).

IDENTIFICATION OF SOME ORGANIC MICROPOLLUTANTS IN URBAN SEWAGE SLUDGES

M. ROCHER, A. COPIN
Faculté des Sciences agronomiques de l'Etat
5800 Gembloux - Belgique

Summary

This work is sponsored by the "Groupe d'Etude pour la Valorisation en Agriculture des Boues résiduaires de stations d'épuration" which joins six research departments of the "Faculté des Sciences agronomiques de L'Etat à Gembloux - Belgique" under the supervision of Mr. R. IMPENS and Mr. J. LECLERCQ.

We would like to thank the "Service de chimie analytique de la F.S.A.Gx" whose collaboration contributed to the achievement of this study.

Depending on the amount of sludges applied on agricultural lands, some compounds present in urban sewage sludges, could alter the quality of the environment and the plant yields.

A general characterization of the organic matter and the identification of some potential micropollutants was performed primarly.

On the other hand the research was orientated in the direction of other compounds as aliphatic and polyaromatic hydrocarbons, glycerides and fatty acids.

1. General characterization of the organic matter of sewage sludges and identification of some potential micropollutants.

The results accumulated during four years on sewage sludge, produced by four water treatment plants in Wallonia (G.E.U.B., Rapports d'activités 1977-1978, 1980-1981), allow to situate the organic matter content (volatil matter at 550°C) of those sludges at approximately 55% of the dry matter.

In respect to its availability to the soil microflora, this organic matter constitutes a favorable asset regarding the agricultural utilization of the sludges.

Yet, we looked into the characterization of this organic matter according to different determinations. Thus, we have investigated the presence in the sludges of indesirable compounds, such as, phenols, anionic detergents, anilines and cyanides. Up to 80 ppm of detergents were detected in the supernatant following sludge centrifugation. The phenols were found in significant amounts in sludges originating from water treatment plant receiving effluents from a food industry. In some cases, we established the presence of total cyanides, however, no such evidence was found in respect to decomposable cyanides.

We also performed the extraction of the sludges using diethyl ether. The ether extract constitutes an important part of dry matter, and in some sludges we found out that it represented 30% of that dry matter.

2. Fatty acids and hydrocarbons characterization of the sewage sludges.

The ponderal importance of that extract has oriented our research towards the characterization of certain other compounds which can be present :
        -the aliphatic and polyaromatic hydrocarbons,
        -the clycerides and fatty acids.

Figure 1 schematizes the analytical steps for the determination of hydrocarbons, glycerides and fatty acids on a sludge sample.

Figure 1 : Analytical steps for characterization of hydrocarbons, glycerides, fatty acids on a sludge sample.

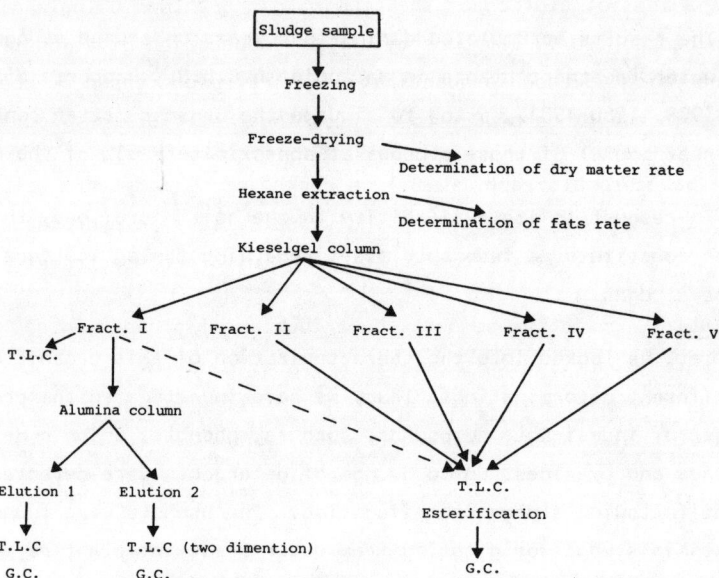

```
                          Sludge sample
                                |
                             Freezing
                                |
                          Freeze-drying ──────────→ Determination of dry matter rate
                                |
                          Hexane extraction ──────→ Determination of fats rate
                                |
                          Kieselgel column
        ┌──────────┬──────────┼──────────┬──────────┐
     Fract. I   Fract. II   Fract. III  Fract. IV   Fract. V
  T.L.C.
        |
   Alumina column
     ┌──────┐
  Elution 1  Elution 2
     |         |
  T.L.C     T.L.C (two dimention)          T.L.C.
  G.C.      G.C.                         Esterification
                                            G.C.
```

The sample is freeze dried. An extraction using hexane is performed on the freeze dried matter. The concentrated extract is submitted to an absorption chromatography on a kieselgel column (Kieselgel Merck 7754 dried at 120°C during 2h and then deactivated with 5% water p/v). Elutions using solvents of increasing polarity permit the separation of the extracted compounds of the sludge into five fractions.

2.1. The hydrocarbons

The hydrocarbons are found in the first fraction resulting from the hexane-benzen (50/50 v/v) eluate on Kieselgel column.

An initial thin-layer chromatography (T.C.L.)(Polygram M.N. SilG) using hexane as a mobile phase, detects the aliphatic and polyaromatic hydrocarbons, as well as their separation with compounds of higher polarity.

A part of that fraction, chromatographed on a neutral alumina micro-column (Merck 1077) activated at 550°C for 5h, permits the separation of aliphatic from polyaromatic hydrocarbons.

2.1.1. The aliphatic hydrocarbons

The initial elution of the alumina micro-column using a mixture of hexane-benzen (90/10 v/v), permits the recuperation of aliphatic hydrocarbons. The analytical method is quantitative for straight chain aliphatic hydrocarbons in the range of C_{14} to C_{26} (COPIN et al 1982).

A gaz chromatography (G.C.) on packed or on capillary column is performed in order to characterize individual aliphatic hydrocarbons. Their presence is subsequently confirmed using infra-red spectrometry.

Table I gives the ponderal distribution of alcanes in a sludge sample of each of the four studied plants.

2.1.1. The polyaromatic hydrocarbons

A second elution with benzen eluates the polyaromatic hydrocarbons (P.A.H.) from the alumina microcolumn. Table II gives the P.A.H. we have investigated.

TABLE I : Ponderal distribution of alcanes in a sludge sample of each of the four studied plants (results in ppm/dry matter)

sludge C_{number}	F_{103}	S_{97}	B_{98}	R_{90}
C15	31,24	27,36	25,90	6,96
C16	43,78	38,77	30,34	7,18
C17	51,64	54,41	55,34	5,24
C18	46,28	51,77	55,09	5,73
C19	35,98	34,46	31,85	12,51
C20	16,56	18,93	26,15	4,71
C21	14,75	15,71	26,63	3,35
C22	13,26	13,97	25,0	6,56
C23	14,95	19,29	41,26	18,34
C24	14,30	21,21	/	19,76
C25	16,49	34,69	64,36	28,29
C26	11,13	20,16	69,88	21,83
	-----	-----	------	------
Total	310,38	350,73	451,80	140,46

TABLE II : Polyaromatic hydrocarbons investigated in sludges

NAME	NUMBER OF CYCLE
Fluoranthen	4
3-4 Benzo-fluoranthen	5
11-12 Benzo-fluoranthen	5
3-4 Benzo-pyren	5
1-12 Benzo-perylen	6
Indeno 1-2-3 cd pyren	6
Phenanthren	3
Chrysen	4
Pyren	4
Perylen	5

A gaz chromatography (G.C.) on packed or on capillary column is performed in order to characterize individual P.A.H. Table III gives retention
times corresponding to different P.A.H. in respect to fluoranthen.

The utilization of a mixte alumina and cellulose thin layer two-dimen-
tional chromatography (Alox Cell AC MIX 25) as discribed by BORNEFF and
KUNTE (1976) confirmed some of the results obtained on G.C.

Therefore, we were able to presume that underlined P.A.H. in tabble II
are present in three plants. However no such presumption was possible concerning the other P.A.H. either because of a lack of sensitivity of T.L.C.,
or because of a difficult separation of two P.A.H. in G.C. or in T.L.C..
Furthermore, individual reference products were unavailable.

The utilization of both liquid phase chromatography and fluorimetry or
the coupling of G.C. and mass spectrometer seem indispensable if research
were to be carried on.

TABLE III : Retention times corresponding to different P.A.H. in respect
to fluoranthen

	C.P.G. classique		C.P.G. capillaire				
	Col.1 Supelco	Col 2 SE 30 10% res.	OV$_1$ ± 15 m	SE 30 ± 10 m	OV101 ± 15 m	GESE54 ± 16 m	SE30 VERSELLE 25 m
Acénaphtène	0,343	0,441	0,214	0,424	0,434	0,533	0,511
Fluorène	0,454	0,542	0,317	0,537	0,538	0,619	0,603
Phénanthène	0,692 0,702	0,741	0,566 -----	0,724 0,735	0,719 0,736	0,826 0,833	0,792 0,802
Fluoranthène	1	1	1	1	1	1	1
Pyrène	1,055	1,042	1,069	1,024	1,030	1,025	1,037
Chrysène	1,403	1,375	1,612	1,314	1,264	1,174	2,019
Benzopyrène a	1,927	-----	2,137	-----	-----	1,345	2,337
Pérylène	1,990	-----	2,168	-----	1,505	1,367	2,356
Benzopérylène ghi	-----	-----	2,593	-----	-----	-----	------

2.2. Glycerides and fatty acids

Following elution of hydrocarbons using hexane-benzen mixture (50/50
v/v), we let successively through the Kieselgel column, benzen for the
desorption of tri-glycerides, a benzen-diethyl ether mixture (90/10 v/v)
which desorbes the di- and mono-glycerides, and finally methanol to desorbe
compounds with higher polarity such as sterols. A T.L.C. (Polygram M.N.
SilG of 25 mm) using a mixture of petroleum ether-diethyl ether-formic acid
(90/60/2,25 v/v/v) as mobile phase, permits to verify the presence of
different classes of the investigated compounds in each fraction.

WATHELET (1978) describes a hydrolysis and esterification method which
permits to chromatograph and dose on gaz chromatograph the methyl esters
products.

The obtained results, show that fatty acids are present in significant
amounts in all sludges investigated. The tri-glycerides are quantitatively
less important than fatty acids. The most frequent and abundant fatty acids
are palmitic, oleic, myristic, stearic, linoleic, linolenic acids.

3. Conclusion

This work should be completed by a pollution risk assessment study of groundwaters. Trials on percolation column submitted to pluviometric conditions of the region under study, are a suitable approach.

Futhermore, the influence of soil microflora on the persistence and behaviour of indesirable compounds should also be observed. We would, therefore, be able to determine the real environmental impact of those sludges originating compounds.

REFERENCES

- BORNEFF J., VON KUNTE Z.F., Wasser und Abwasser Forchung, (1976),9,35.
- COPIN A., DELEU R., MOUTEAU A., ROCHER M., The Sciences of Total Environment, (1982), 22, 179-187.
- G.E.U.B., Groupe d'Etude pour la valorisation des Boues, Rapport d'activité, 1977-1978, Rapport d'activité 1979, Rapport d'activité 1980-1981, Faculté des Sciences agronomiques de l'Etat - B5800 Gembloux - Belgique.
- WATHELET J.P. (1978), Contribution à la determination de la structure des triglycerides - Etude d'huiles et de graisses entières ou fractionnées, Thèse de Doctorat, Faculté des Sciences agronomiques de l'Etat - B 5800 Gembloux - Belgique.

PRESENTATION OF THE ANALYTICAL AND SAMPLING METHODS AND OF RESULTS ON ORGANO-CHLORINES IN SOILS IMPROVED WITH SEWAGE SLUDGES AND COMPOST

P. DIERCXSENS & J. TARRADELLAS

Swiss Federal Institute of Technology of Lausanne

Institut du Genie de l'Environnement

EPFL-Ecublens

1015 Lausanne (SWITZERLAND)

SUMMARY

This paper is a synthesis of various works realised the two last years in our laboratory on sewage sludges and composts.
First, the PCB concentrations in the sewage sludges of nine water treatment plants of Switzerland are given and discussed.
The relation between compost and sewage sludges are underlined in a second short chapter.Then the analysis of PCBs in soils improved with sewage sludges and compost are discussed, as well as those in the earthworms.
Earthworms can constitute an excellent indicator of soil pollution by micro-pollutants and PCBs in particular.
The analysis of PCBs from earthworms are realised by acid digestion followed by n-hexan extraction, sulfuric acid and Florisil column cleanup and glass-capillary-gas-chromatography with ECD detection (split mode, methyl-phenyl-silicone SE 54). A tipical chromatogram is shown.
This study has shown that earthworms are better punctual indicators of the average PCB contamination of soils than a soil sample.

1. INTRODUCTION

This report is a synthesis of various works realised the two last years in our laboratory on sewage sludges and composts.
It is subdivided in four parts :
- the first one presents some results of PCB analysis on the sewage sludges of nine water treatment plants and discusses them.
- the second one deals with the PCB analysis in compost. Those composts are partially produced by means of sewage sludges.
- the third one discusses the PCB analysis in soil samples improved with sewage sludges and compost.
- the last part is more analytical and presents certain details of the analytical procedures adopted.

2. SOME RESULTS OF PCB ANALYSIS IN SEWAGE SLUDGES (1)

Four types of water treatment plants were chosen :
1) three water treatment plants of big cities:
- Zürich
- Genève
- Lausanne
2) three water treatment plants of industrial cities:
- Baden
- Winterthur
- La-Chaux-de-Fonds
3) 1 water treatment plant of a little town:
- Morges
4) two water treatment plants of industries:
- Papier Fabrik Utzenstorf
- Fairtec AG

The concentrations of PCBs in the sludges were established (TABLE 1)
- in the sludges with 95,5% water
- in the sludge cake after dehydration (it is in fact in this form that the sludges were sampled, except in Baden where there is no dehydration plant)
- on dry material after liophylisation.

comments

- All the sewage sludges analysed contained PCBs
- It is noticeable that it is not justified, without a better knowledge of the PCB distribution in a water treatment plant, to pretend that all the PCBs remains in the sludges. This idea is cosolidated by a study performed by Aswald & all. (2) in our laboratory. This study illustrates the geographic distribution of the PCB concentrations in the sediments of the

Water treatment plant	Inhabitant equivalent (1979)	Estimation of produced sludges (95,5% H₂O) (metric tons)	Concentration of PCBs in the sludges (ppm)			Extrapolation annual PCB quantity (Kg/year)
			Sludges with 95,5% water	Sludge cake	Dry material	
Zürich	345.000	235.000	0.130	1.80	2.80	31.0
Genève	268.000	183.000	0.340	2.90	7.60	62.0
Lausanne	209.000	143.000	0.038	0.48	0.85	5.4
Baden	48.000	33.000	0.110		2.40	3.6
Winterthur	100.000	68.000	0.045	0.30	1.00	3.1
La-Chaux-de-Fonds	36.000	25.000	0.016	0.14	0.36	0.4
Morges	20.000	14.000	0.054	0.36	1.20	0.8
Papier fabrik Utzenstorf (BE)	43.000	30.000	0.110	0.85	2.50	3.3
Fairtec (AG)			0.022	0.21	0.48	

TABLE I : PCB CONCENTRATION IN THE SEWAGE SLUDGES OF NINE WATER TREATMENT PLANTS OF SWITZERLAND

Leman lake. The concentration is maximum on the outskirts of big cities, like Geneva, Lausanne and Montreux. This fact seems to indicate that a part of the organo-chlorines could be rejected in water treatment plant effluents. The distribution of PCBs in water treatment plants is now a study subject in our laboratory.

- By way of comparison we can cite :
  - the study realised on a Sweden water treatment plant by Mattson and Nygren (3), which shows us on 5 analysis a mean concentration of 0.9 ppm on a dry weight basis.
  - analysis made in 1982 in our laboratory for preparing the COST inter-calibration on PCBs (4) gave the following results on liophylised sewage sludges:
    - middle town in the north of Italy   5.9 ppm
    - big city of the south of Germany :
                     residential district   5.4 ppm
                     industrial district   6.5 ppm
    - sludges coming from domestic waters
      of a city of the north of Germany   3.4 ppm
  - the articles written by Alun E. & all. on the distribution of poly-chlorinated biphenyls in sewage sluges (5,6)
  - with a vieuw to fix the ideas, without that it have any comparison value, the average mercury content of 68 Swiss water treatment plants was 5 ppm on a dry weight basis.
- We observe that our results are situated in the same concentration limits than those cited as comparison.

3. SOME RESULTS OF PCB ANALYSIS IN COMPOST

A research, realised in 1980 in our laboratory (7), into PCB contents in compost revealed the following results:
- like the sewage sludges, all the compost samples showed the presence of PCBs.
- All the compost samples analysed presented concentrations higher as 1 ppm (fresh weight), which must be considered as important.
  The results were situated between 1 and 3.5 ppm on a fresh weight basis and between 1.8 and 6.9 on a dry weight basis.
  The samples were coming as well from water treatment plants of industrial cities, non industrial cities and rural countries.
  The PCB concentrations were all of the same order of magnitude

- Much of these composts are mixtures of household refuse and sewage sludges. On first approach, it seems not justified to attribuate the presence of PCBs in compost uniquely to sewage sludges. Two reasons about this:
  . a compost exclusively formed with household refuse, coming from the plant of Uvrier in central Valais, contained up to 2.67 ppm of PCBs (fresh weight)
  . in the compost plant of Penthaz (collecting the household refuse of the north of Lausanne), the sewage sludges added to the household refuse had a low PCB concentration (0.19 ppm fresh weight), which do not prevent that the compost of this plant had one of the highest PCB concentration analysed (3.84 ppm fresh weight).

4. ANALYSIS OF PCBs IN SOILS IMPROVED WITH SEWAGE SLUDGES AND COMPOST

The third part of this report is the study of the PCB concentration in a soil improved with sewage sludges and composts and in the earthworms living in it. The sampling site is a vineyard of the Lavaux, situated near the village of Chardonne (Vaud). The slope of the vineyard is 10 to 15% inclinated, the leaching is then important. For that reason the compost and the sewage sludges have not only an importance as fertilizing agent but as physical stabilizing too.

the_soil_analysis_difficulties

A soil presents with regard to micropollutant analysis two principal disadvantages:
- the first is purely analytical. The sand and several other soil components obstructs easily the glassware utilised or hinders the impervious closing of it. We must frequently have recourse to complicated methods for recover the hexanic phase containing the micropollutants. Moreover there is an obvious loss of information during this difficult recovering.
- the second is from the obtaind results point of vieuw. We have observed a great heterogeneity in the analysis results of the soil.
  One of the reasons is the following. The PCBs are essentially adsorbed on the organic and clayey particles of the soil (8,9). Certain microscopic fat nodules brought eighter by the sewage sludges, eighter by agricultural engines are punctually distributed on the parcel.
This two problems have brought us to consider the soil analysis on an other point of vieuw. Certain animals, living in narrow contact with the soil, could furnish us interesting results. We have thought to study the earthworms as information source about soil contamination.

interest of the analysis of earthworms

A good exemple of illustrating the state of contamination of a terrestrial ecosystem is the earthworm. That is:

. they represent the first animal biomass of dry land (10,11,12).
  This represents an analytical advantage, in so far as it is relatively easy to obtain a great quantity of them.

. they remain always in the ground.

. one can find them everywhere

. they occupy an intermediate place between the microorganisms and the superior animals in the trophic relations (13).

. they ingest enormous quantities of earth: in 1 hectare of land, up to 300 metric tons would be ingested (10).

. the assimilation membranes of the worms are comparable with those of the major part of the other animals, including those of superior animals, but contrary to the assimilation membranes of the plants. This gives them a remarkable bio-sampler value.

Earthworms represent, therefore, a good exemple in the study of micropollutant fluxes (15,16,17,18). Moreover, the global study of the contamination of the agricultural ecosystem by sewage sluges (19,20,21,22,6,9) cannot be performed without the knowledge of the PCB concentration in the earthworms. That is: . earthworms constitute an integration of pollutant rejection during the time.

   . earthworms constitute the food of a great number of mammals and birds.

   . an eventual toxicity on the worms would in due time induce a decrease in soil fertility.

some results

The following results attempt to demonstrate the differences in PCB concentrations observed in the soil of the vineyard and in the worms living in it. A mixture of compost and sewage sludges of 0.9 ppm (fresh weight) has been spread 4 years before the sampling of the worms and the earth.
The PCB content of the compost was estimated by means of an analysis made in 1980 on the compost produced in the same treatment plant.

	results (ppm fresh weight)	divergence to the mean in %
mean concentration of PCB in the tissues	0.65 ± 0.05	8.5
mean concentration of PCB in the intestinal earth	0.44 ± 0.05	12.0
mean concentration of PCB in 3 soil samples	0.08 ± 0.05	60.0

TABLE II: Differences in the PCB concentrations in the
earthworms's tissues, in the intestinal earth
and in the earth in which they feed.

Two important comments after reading this results:

- First, the difference of concentration between intestinal earth and soil
  samples seems to indicate that the worms makes a choice of earth they
  ingest. We may validly propose this hypothesis owing to the fact that
  the soil of the vineyard where those results were established was completely free from litter, which could constitute an addition of pollutants other than that of the earth.

- Second, it is noticeable, on TABLE II, that the relative dispersion of
  the soil results, for a piece of ground as small as that where we did
  our samples (1000 m^2) is much higher than that of the earthworms sampled
  at the same place.
  This seems to indicate that earthworms would be better punctual indicators of the average PCB contamination of the soil than soil samples.

The different methods for emptying the digestive tract of the earthworms
are largely discussed in our previous work, i.e.:"Methods of extraction and
analysis of PCBs from earthworms" (23).

soil-plant relation (9,24)

The extent to which PCBs and other materials are absorbed by plants and may
be toxic to human beings through food crops is not well understood.
Iwata and Gunther (25) found that PCBs were translocated from the soil into
carrots. The peel contained 97% of the root residue and very little was
found in the foliage.
FDA (26) indicated that sludges containing 10 mg/l PCBs on a dry weight
basis should not be used for growing food crops.
Environment Canada (22) detected PCBs in some tissues of mixed grain and corn.

The level which were detected in the tissues of one sample of mixed grain was 28 ppb (soil: 10 ppb) and for two corn samples 5 and 95 ppb respectively (soil: 15 and 47 ppb). Although, the ultimate fate of PCBs disposed of as landfill or as a soil supplement has not been sufficiently studied, sludges have been used for agricultural production and have shown to result in up-take of some PCBs in crops. Thus, it may be concluded that there is a need to establish regulations controlling the disposal of such wastes in agriculture.

5. ANANLYTICAL ASPECTS

Need of the use of a capillary column.

The figure I shows a tipical chromatogram obtained after preparation of the earthworms in an artificial medium, called "Artisol" (23). With marked stan-darts of PCB, the capillary column distinguishes between PCBs and other or-ganochlorines now found in the soils.

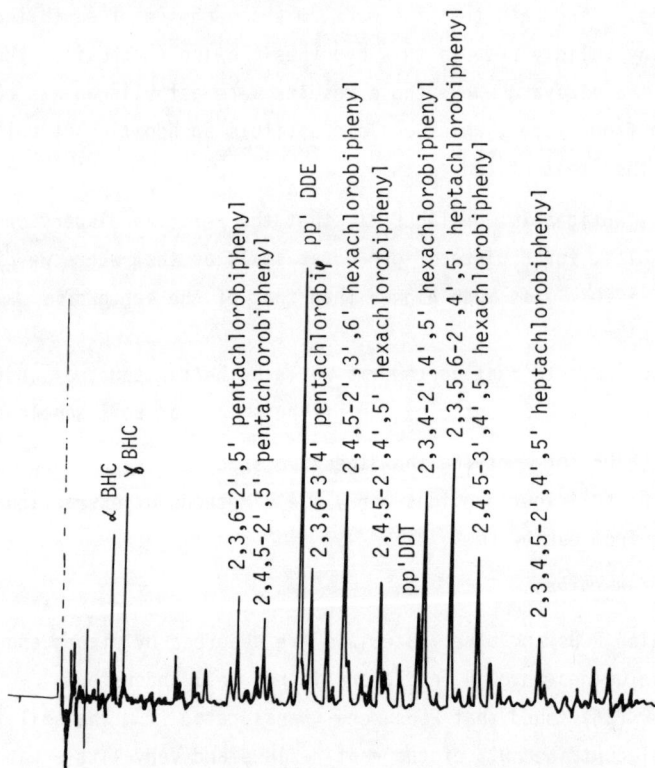

FIGURE I : CHROMATOGRAM OF PCBs FROM EARTHWORMS
TISSUES

CONCLUSIONS

The several analysis indicated that sewage sludges from industrial and municipal wastewater treatment plants invariably contained PCBs. In a majority of
the sludges the PCB were present in a form resembling Aroclor 1254 and 1260.
There is a need to establish regulations controlling the disposal of such
wastes in Agriculture.Clearly, further studies on the occurence and significance of polychlorinated biphenyls in the environment will be required
prior to the establishment of specific regulations on PCBs in relation to
agricultural production.

It would be of interest to extend this study to other organochlorines like
PCDD, PCDF, which have been found in some municipal sludges in concentrations
up to 300 ppt (14). Occurence of PCT, PCN and chlorinated paraffins are
also of interest.

REFERENCES

1. Symposium "Les PCB en Suisse", Ecole Polytechnique Federale de Lausanne,
   Institut du Génie de l'environnement pp.62 (april 1980)
2. "Etude préliminaire de la pollution du Léman par les PCB"
   7th Postgraduate Course in Environmental Engineering of Swiss Federal
   Institute of Technology of Lausanne. Internal report pp.71 (nov. 1979)
3. P.E. Mattson & Nygrens, J. Chromatogr.,124,pp.265-275 (1976)
4. J. Tarradellas:"Report on the preliminary analysis of sludges, soils
   and sediments to detemine the most adequates for a COST intercalibration
   about the PCB concentration of this samples", EPF-Lausanne (march 1982)
5. A.E. Mc Intyre, R.Perry & J.N. Lester, Environ. Pollut. Ser.B,2,
   pp.223-233 (1981)
6. A.E. Mc Intyre, J.N. Lester & R. Perry, Environ. Pollut. Ser.B,2,
   pp.309-320 (1981)à
7. O.Sow:"Etude preliminaire de la contamination des composts par les PCB"
   8th Postgraduate Course in Environmental Engineering of Swiss Federal
   Institute of Technology of Lausanne. Internal report (nov. 1980)
8. Haque & all.,Env.sc.techn.,8,pp.139 (1974)
9. M. Suzuki, Arch. env. cont & tox.,5,pp.343 (1977)
10. M.B. Bouché : Lombriciens de France : écologie et systematique
    (I.N.R.A. Publ. 72-2, Paris) 671 pp. (1972)
11. A. Stöckli, Landw. Jahrb. Schweiz., 72,pp.699-725 (1959)
12. E. Nowak, Pol. Ecol. Stud.,2,4,pp. 195-207 (1976)
13. M.B. Bouché :"La vie dans les sols (Gauthier Villars,Paris),pp.187-209

(1971)

14. I.L. Lamparski & T.J. Nestrick:"The isomer specific determination of TCDD
     at part per trillion concentrations in : "Chlorinated dioxins and rela-
     ted compounds, impact on the environment." Pergamon Press pp.1-14 (1982)

15. D.V. Yadav, M.K.K. Pillai & H.C. Argawal, Bull. Environm. Contam. Toxi.
     col.,16,5,pp.541 (1976)

16. B.N.K. Davis & R.B. Harrison,Nature,211,pp.1424 (1966)

17. G.A. Wheatley and J.A. Hardman, J. Sci. Food Agr. 19,219 (1968)

18. B.N.K Davis & M.C. French, Soil Biol. Biochem. 1,45 (1969)

19. A.K. Bergh & R.S. Peoples, Sci. Total. Environ.,3,pp.197-204 (1977)

20. C.Schweizer and J.Tarradellas, Chimia,34,pp.509-519 (1980)

21. D. Pal., J.B. Weber & M.R. Overcash, Residue Revieuw, 74,pp.45 (1980)

22. Environment Canada, Techn. Report, 76-1,pp.61-65 (1976)

23. J. Tarradellas, P. Diercxsens, M.B. Bouché
     "Methods of extraction and analysis of PCBs from earthworms"
     12th Annual Symposium on the Analytical Chemistry of Pollutants
     Amsterdam (april 1982) (in press)

24. E.Epstein & R.L. Chaney, Journal WPCF,pp.2037-2042 (1978)

25. Y. Iwata, F.A. Gunther, Arch. Environ. Contam. Toxic,4,44 (1976)

26. C.F. Jelinek: "Management of sludge use on land, FDA Considerations"
     Paper presented at conference in connection with the International
     Water Conservancy Exhibition (Sept. 1-5-1979)
     Copy available from U.S. EPA. Cincinnati, Ohio

DISCUSSION ON SESSION I: EFFECTS OF ORGANIC MICROPOLLUTANTS

Chairman: T.W.G. HUCKER - Rapporteur: J.E. HALL

(After papers by Lester, Lindsay, Bridle and Webber)

Chairman

I would like to ask Dr Webber whether he observed any phytotoxic effects
on swiss chard which could be attributed to organic contaminants in the sludge.

M Webber (Canada)

Effects on plant growth appeared to be attributable more to salt concentrations than organics. The salt content of the sludge-soil mixtures was well into
the range of electrical conductivity capable of affecting plant growth. It is
intended to analyse the plant material for organics although this poses considerable analytical problems.

J Tjell (Denmark)

Did Dr Webber consider aerial inputs of organics?

M Webber

There was a probability of airborne contamination in this trial which was
conducted near a steel plant. This could explain negative effects on the control.

G Fleming (Ireland)

I would like to support Dr Lindsay's observations concerning direct
ingestion of soil by grazing animals. Soil can account for 6 per cent of the
dry matter intake of sheep or 400 g per day. It can therefore be an important
route for contaminants entering the diet of grazing animals and must always be
considered.

D Miller (UK)

For estimating recovery of organic contaminants in the analysis of sludge,
Dr Bridle had enriched sludge directly with the contaminants concerned. In
work with metals it was realised that this could lead to spurious results which
could be avoided by enriching sewage and then preparing the sludge from the
enriched sewage. Was Dr Bridle aware of this potential problem and was it
influencing his results?

T Bridle (Canada)

Dr Miller's point is valid but I would like to make it clear that for
toxicity investigations we are using only sludge high in indigenous organics.
The spiked sludge is used just to check recoveries on analysis. I don't know
whether recovery would be the same from spiked sludge as from sludge containing
the same concentration of indigenous contaminant.

After papers by Sauerbeck, Rocher and Diercxsens:

J Tjell

I would like to broaden the discussion and come to some conclusion as to whether or not organics in sludge are a widespread problem.

D Lindsay (UK)

I suggest that problems with organics are likely to be localised to sludges from sewage treatment works receiving particular kinds of industrial effluent. It is impossible to say what level of contamination is safe or unsafe. A compromise has to be struck between the need to dispose of sludge and protection of the foodchain. Unlike cadmium, there is no firm evidence of translocation of organic pollutants into crops.

Chairman

We are still in the position where accurate means of analysis of organics are neded before the problem can be properly assessed. It is probably too early for limits. We need more information from monitoring on which to assess the position.

J Lester (UK)

Most of the information available is 10-20 years old. Therefore we cannot present an accurate picture of the current position. More monitoring is needed, especially bearing in mind that the amounts and number of organic pollutants are continually increasing.

D Miller

This raises the question of what is meant by monitoring. Is it in reference to routine operations or research work? The water industry would be reluctant to take on widespread routine monitoring for organics as well as for heavy metals.

J Lester

I am not suggesting a blanket approach to monitoring. If the use of a new organic pollutant were notified, the monitoring could be localised.

D Lindsay

All new chemicals will have to be tested for environmental effects before their use so we should be able to control them. The problem now is with existing chemicals.

Chairman

Are there any limits for organics in Canada?

M Webber

The Ontario (Canada) guidelines recognise the potential organic problem but without specifying limits.

R Davis (UK)

What is the basis for selecting test crops for experiments on organics? Why was Swiss chard chosen?

M Webber

It was selected because it has been widely used for metal work being a metal accumulator, although it is not known to accumulate organics. There is some evidence that carrots accumulate PCBs.

D Lindsay

There is a suggestion in the literature that oilseed crops may take up organics comparatively readily. High levels of PCBs observed in carrot could be due to surface contamination.

SESSION II - EFFECTS ON INORGANIC MICROPOLLUTANTS

Limits of zinc and copper toxicity from digested sludge applied
to agricultural land

Zinc, copper and nickel: suggested safe limits in sewage sludge
treated soils

Sludge application to land: overview

The intake by man of cadmium from sludged land

Total and biorelevant heavy metal contents and their usefulness
in establishing limiting values in soils

Cadmium concentrations in field and vegetable crops - A recommended
maximum cadmium loading to agricultural soils

Cadmium in sludge-treated soil in relation to potential human
dietary intake of cadmium

Soil-chemical evaluation of different extractants for heavy metals
in soils

Effects of cadmium humates prepared at pH4 and pH6 on the growth
and mineral composition of maize plantlets cultivated in nutrient
solutions

Tolerable amounts of heavy metals in soils and their accumulation
in plants

Plant uptake of cadmium: summary of Swedish investigations up
to and including 1981

Cadmium speciation in soil solutions of sewage sluge amended soils

Influence of different types of natural organic matter on the
solubility of heavy metals in soils

Possibilities of reducing plant availability of heavy metals
in a contaminated soil

Geochemical pollution - Some effects on the selenium and
molybdenum contents of crops

Effects of heavy metals on soil microorganisms

Heavy metals in soils, sludges and plants: summary of research
activities

Discussion

LIMITS OF ZINC AND COPPER TOXICITY FROM DIGESTED SLUDGE APPLIED TO AGRICULTURAL LAND

N.B. JOHNSTON*, P.H.T. BECKETT* and C.J. WATERS[ø]

* Department of Forestry & Agricultural Science, Oxford University

[ø] Directorate of Scientific Services, Thames Water, London

Summary

Trial plots were constructed on a sandy clay loam soil to examine the effects of a range of dressings of high zinc liquid digested sludge on yields and metal content of barley, ryegrass and lettuce grown on the plots. The heaviest dressing contributed 200 $kgCuha^{-1}$ and 1600 $kgZnha^{-1}$ in one application. In the next three years barley and ryegrass on the plots with most sludge showed severe "lodging" due to excessive N, and also some depression of germination. These are attributable to excessive levels of mineralisable organic nitrogen and decomposable organic matter in the heavy sludge applications, far in excess of any normal practice. There was little or no evidence of zinc or copper toxicity; it appears to require at least four times the quantity of sludge usually applied over 30 years, to achieve phyto-toxicity in these crops.

1. INTRODUCTION

We are concerned here with immediate phyto-toxicity. The following
trials were planned to determine just how much heavy metals may be applied
in sludge to a typical soil without inducing toxic reactions in barley,
ryegrass or lettuce crops in the years immediately following.

2. EXPERIMENTAL

The soil was a freely-drained sandy clay loam of the Sutton series,
typical of good agricultural land in lowland Britain. Initially at pH 5.2
the soil was limed to pH 6.5 and brought to good Mg, K and P levels for the
duration of the trials.

Sixteen trial plots (each 10 m^2) were treated in 1978 (Year 1) with
liquid digested sludge in quantities equivalent to 0 - 5 times the then
recommended maximum 30 year dressing.[1,2] Each plot was surrounded by an
impermeable plastic membrane down to 60 cm below the soil surface to prevent
contamination between plots. The composition of the sludge varied so much
even within one tanker-load that it proved impossible to achieve replicate
treatments. Table 1 lists the quantities applied, which will be referred to
as multiples of S, the maximum zinc equivalent in 30 years as then recommended. In addition one set of similar plots received graded amounts of
inorganic N and P fertilizer (up to 700 and 300 kgha^{-1} respectively) in each
of 1979, 1980, 1981 (Years 2-4).

Subplots of eight sludge treatments (Table 1) were cropped to
perennial ryegrass and to lettuce in years 2-4: the mature lettuce was
harvested in the usual way, and the ryegrass was cut two or three times each
year. All the inorganic plots and two subplots on each sludge treatment
were cropped to barley (spring barley in year 2; winter barley in years 3
and 4). Eight subplots were cropped at the five-leaf stage and the others
at maturity; and at various stages of growth the dry matter yield of barley
was assessed non-destructively by MMB grassmeter. Crops were weighed fresh
and dry, and analysed for N, P, Zn, Cu and Ni by usual methods.

Full details and results of the trial, and of other plots and subplots
not reported here, are to be published (Johnston, in preparation).

3. RESULTS

The sludge used was relatively low in Ni so the Ni results are
omitted. The data of yields and tissue composition were subjected to the
usual statistical examination and Figs. 1-4 present all the effects of the
sludge, or fertilizer N treatments, that were found to be significant. Non-

significant results are presented as means with 95% confidence limits. 100 kgNha^{-1} is approximately equivalent to 1 S. The lettuce yields appeared to be depressed by the 3.7 S sludge treatments, but the trend was shown to be not significant.

4. DISCUSSION

Note that to achieve large application of Cu, Ni and Zn, required very high sludge applications indeed. Inevitably these produced effects, particularly inhibition of germination and delay in the establishment of young crops, which were nothing to do with their heavy metals, and would not have occured had the quantity of sludge been spread at the recommended rate (ie spread over 130 years).

Clearly the sludge raised tissue Cu and tissue Zn (Figs. 1, 3 and 4), more so in the young than in the mature crops.

Equally clearly barley yields were depressed by very high sludge applications (Fig. 1). At first sight this effect might be attributed to Cu or Zn toxicity. However, the high-sludge plots showed severe "lodging", and the grassmeter demonstrated that the yield of dry matter in the growing barley was not depressed, even on the high-sludge plots, before the onset of "lodging". Furthermore barley in the inorganic N, P plots showed very similar N uptake (not reported here) and almost identical yield curves (Fig. 2) in which the reduced yields of high-N crops could only be due to lodging. Also the tissue Cu and Zn on Fig. 1 lie below their generally agreed toxic levels, and the more sensitive 5-leaf barley with higher Cu and Zn levels (Fig. 3) showed no yield depression at higher sludge levels. It seems clear therefore that the mature barley (Fig. 1) was poisoned by N, not by Cu or Zn, though tissue Cu and Zn were approaching toxicity at the highest sludge treatments.

The ryegrass results are broadly similar (Fig. 4). The grassmeter confirmed that yields on the high-sludge plots were depressed by lodging, and that the first cuts were the most affected.

5. CONCLUSION

On a normal arable soil, with no unusual capacity for fixing heavy metals, single sludge additions of up to 4.3 times the recommended ADAS-10 rate failed to cause phyto-toxicity in barley or ryegrass, but may just have done so in lettuce. These trials demonstrate the risk that the very large quantities of organic matter that must be added in sludge dressings to reach phytoxic heavy metal concentrations may confuse the results. It is possible

that some earlier published sludge trials may have attributed the negative
effects of similarly high additions of decomposable organic matter (which
would not occur in normal practice) to heavy metals. Extreme care is
therefore needed in the design and implementation of practical trials to
assess the effects of sludge additions to land, particularly if the results
of field trials are to be used in establishing limits for the safe disposal
of sludge to land.

6. ACKNOWLEDGEMENTS

     This paper is presented with the permission of Dr M C Dart, Director
of Scientific Services, Thames Water Authority. The authors wish to thank
all of those colleagues who took part in the collaborative project. In
particular they wish to express their gratitude for the cooperation
extended by Mr V H Lewin, Mr D Redhead and Mr D Cornwell. The views
expressed in the paper are those of the authors and not necessarily those
of Thames Water.

7. REFERENCES

     1. Agricultural Development Advisory Service, Permissible Levels of
         Toxic Metals in Sewage used on Agricultural Land. Advisory
         Paper No. 10, 1971.

     2. DOE/NWC Standing Committee on the Disposal of Sewage Sludge.
         Report of the Working Party on the Disposal of Sewage Sludge
         to Land. STC Report No. 5, August 1977.

TABLE 1   PRINCIPAL TREATMENTS OF 16 TRIAL PLOTS

Two sub-plots barley*, one each of lettuce, ryegrass				Barley on two sub-plots each			
DM	Cu	Zn	S	DM	Cu	Zn	S
O			O	O			O
23	14	107	.3	O			O
49	31	232	.7	13	15	105	.3
66	40	299	.8	34	22	157	.4
77	48	361	1.0	31	39	281	.8
142	94	707	1.9	1O1	65	469	1.3
219	135	1020	2.8	183	11O	815	2.2
326	213	1610	4.4	31O	139	1350	3.7

DM = Sludge dry matter tha^{-1}

Cu and Zn in kg metal ha^{-1}

S = Sludge level as multiple of "ADAS-1O" 3O year limit

* one sub-plot harvested at five-leaf stage.

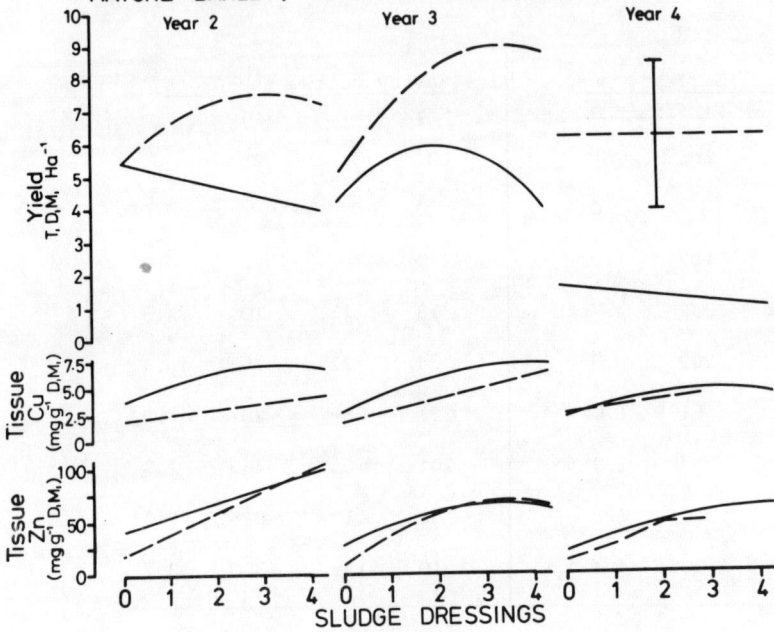

SLUDGE PLOTS
MATURE BARLEY, GRAIN (——) AND STRAW (— —)

FIGURE 1

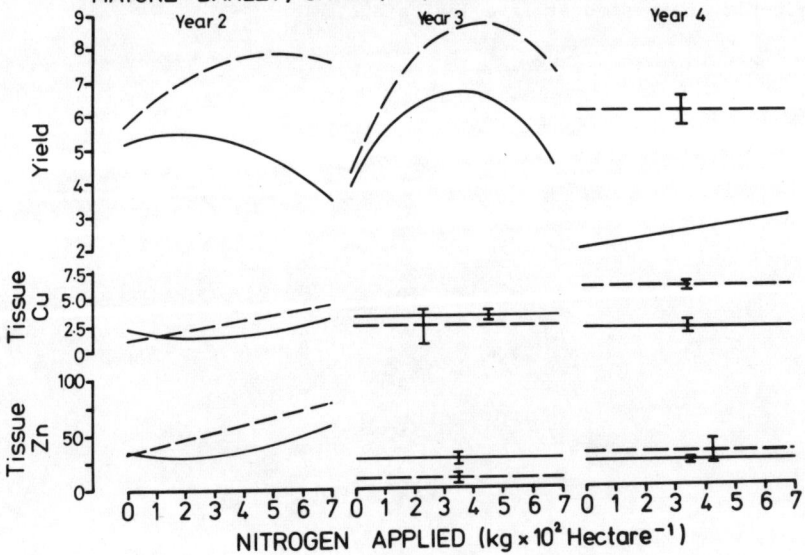

INORGANIC N and P PLOTS
MATURE BARLEY, GRAIN (——) AND STRAW (— —)

FIGURE 2

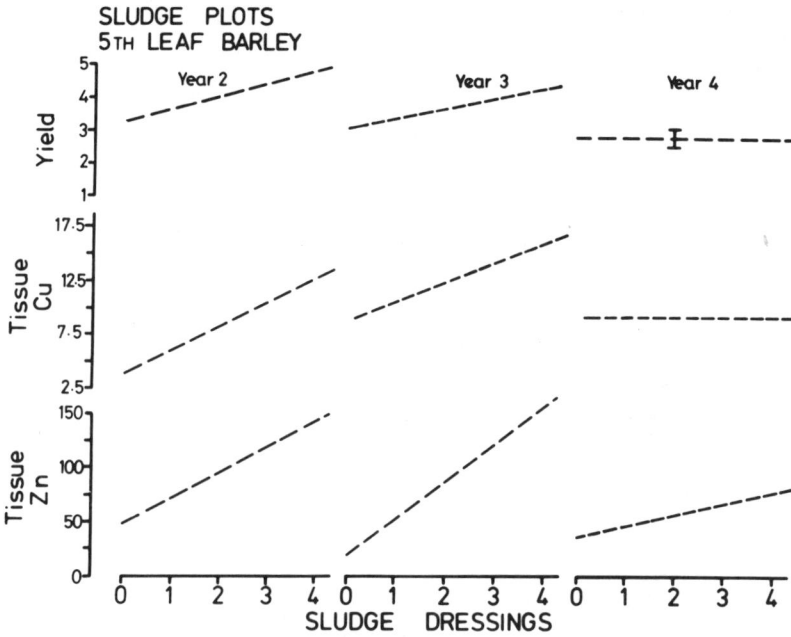

SLUDGE PLOTS
5TH LEAF BARLEY

FIGURE 3

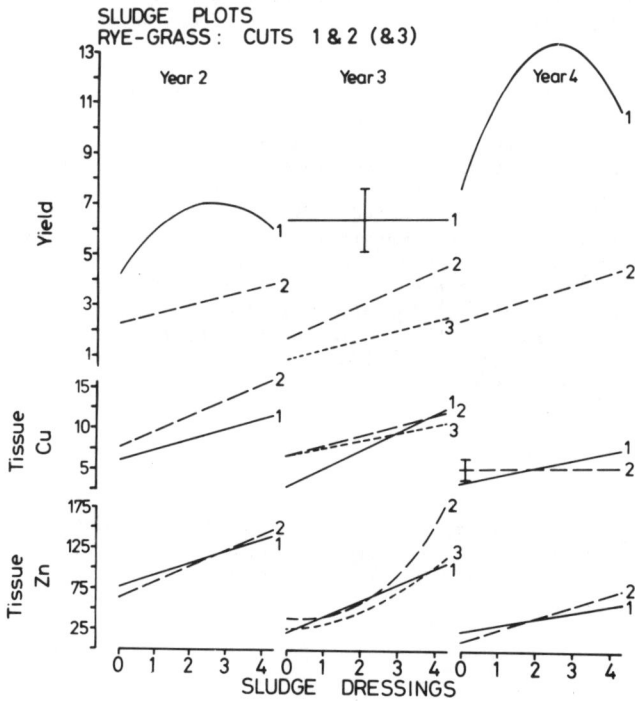

SLUDGE PLOTS
RYE-GRASS: CUTS 1 & 2 (& 3)

FIGURE 4

ZINC, COPPER AND NICKEL - SUGGESTED SAFE LIMITS IN SEWAGE SLUDGE TREATED SOILS

J.H. WILLIAMS

Ministry of Agriculture, Fisheries and Food, Woodthorne, Wolverhampton,
West Midlands, United Kingdom.

Summary

Soil and plant variables which influence the phytotoxicity of zinc, copper
and nickel to crop plants are discussed. Relative toxicity is variable
depending on the crop grown, soil type, soil reaction and on the basis of
the assessment. A method of assessing the quantities of any particular
sludge that may safely be applied over a long period is described. It is
based on a calculated Zinc Equivalent for a particular sludge with a
recommended maximum application of 560 kg/ha of Zinc Equivalent over a
long period. Subsequently it is suggested that sludge metal additions are
checked at intervals by monitoring the increase in the total metal concentrations in the soil. Recommended safe maximum total concentrations of
zinc, copper and nickel in sludge treated soils have been deduced from the
minimum extractable levels of these metals known to be harmful to plant
growth. Metal additions may be determined by subtracting the normal background soil concentration from the measured total soil metal concentration.
In situations where metals of a different origin are present, eg minespoil,
it is necessary to determine the EDTA extractable metal content to ensure
that the zinc, copper and nickel concentrations are well below the safe
toxic thresholds before any sludge is applied.

1. INTRODUCTION

Liquid digested sludge is a useful source of nitrogen and phosphorus for crop plants. Its application to land should be timed to gain the maximum benefit from the nitrogen which it contains and according to the requirement of the crop. Dewatered sludges possess some nitrogenous value but are particularly useful as an aid to maintaining the phosphorus status of soils and as a source of organic matter for poorly structured or unstable soils. However, sewage sludges often contain amounts of toxic elements which can affect crop plants, human and animal health. Metals added to the soil are firmly held by clay and organic matter and, once applied, remain in soil almost indefinitely. In general, crop phytotoxicities are caused by elevated concentrations of zinc, copper or nickel in soils with an acid pH, usually affecting sensitive crops such as vegetables. Problems will be minimal if an upper limit is placed on sludge metal addition and soil pH is maintained at 6.5 or slightly higher.

Several plant variables such as species, cultivar, stage of maturity and plant part affect metal uptake. Most plant species usually respond linearly to soil concentrations of zinc and nickel whereas the plant concentrations of copper show a non-linear response to the application of sludge-borne copper to soil. The availability of these metals to plants after the application of sewage sludge to soil is also dependent on several soil factors including metal addition, pH, cation exchange capacity, organic matter content and the presence of hydrous oxides of Fe and Al. Of these, soil pH appears to be the most critical factor in controlling metal uptake and extent of crop damage. Annual variations are also found in the metal content of crops even if the soil and plant factors remain constant. Temperature and moisture stresses imposed during the growing season are very largely responsible for variations from year to year. In general, metal concentrations are increased if the stress results in decreased dry matter production assuming uptake remains relatively constant.

2. CROP PHYTOTOXICITY

The phytotoxic metal usually present in greatest quantity in sewage sludge is zinc, followed by copper and nickel. Numerous instances of damage to crops from sewage sludge applications in which the zinc extractable from soil by 0.5 M acetic acid was greater than 100 mg/kg have been recorded. Some examples of the effects of extractable zinc levels in soil on crop growth are given in Table I.

Like zinc, the toxicity of copper is also influenced by soil pH and organic matter content. Copper would appear to be adsorbed strongly on to organic matter since large applications of copper contaminated sludges have not given any significant increases in copper uptake by cereals and vegetable crops. It is less readily translocated within above ground plant parts but it accumulates in roots. In cereals for example, it can produce multiple, branched, swollen roots. In a series of microplots at the Wolverhampton ADAS Regional Centre, zinc and copper sulphates were added to a sandy loam soil over the period 1948-50 and a range of crops grown up to 1961 without further metal addition. The rates went up to 800 mg/kg of added zinc and copper in soil. The figures for extractable zinc and copper obtained in 1961 using 0.5 M acetic acid are shown in Table II. The soil pH was 6.5 and toxic effects were noted in crops grown on those plots where the extractable metal content was 150 mg/kg for Zn and 80 mg/kg for Cu.

In a pot experiment at the ADAS Laboratory, Wye a sludge containing 8000 mg/kg of total copper was mixed with soil at rates equivalent to 0, 25, 50 and 75 tonnes/ha to give soil copper concentrations of 0, 80, 160 and 240 mg/kg. The yields of red beet dry matter obtained are shown in Table III. A total copper concentration of 80 mg/kg in soil applied in sludge form resulted in a large reduction in yield of leaves and roots.

Compared to zinc and copper, nickel is injurious to crop growth at much lower soil concentrations although plant species vary appreciably in their sensitivity. Pot experimental work at the Wye and Wolverhampton ADAS Laboratories demonstrated the modifying effect of soil pH on the toxicity of nickel to oats and mustard. The effects of nickel were very toxic at pH 5.7 at only 50 mg/kg added. At pH 6.4 the yield of oats were still depressed at 50 mg/kg added but mustard was unaffected up to 100 mg/kg added. These yield depressions were associated with levels of 12.5 - 15.0 mg/kg of 0.50 M acetic acid extractable nickel.

Greenhouse studies using sewage sludge amended with metal salts have shown that the relative toxicity is very dependent on the crop grown and on soil pH. For example, at a soil pH of 5.7, the relative toxicity of Zn : Cu : Ni was 1 : 4 : 6 and 1 : 1 : 2 for wheat and lettuce respectively (4). for rye, the relative toxicity of Zn : Cu : Ni was 1.0 : 1.8 : 1.0 (1).

Relative toxicity of the metals depends on whether it is based on metal additions or on the soil metal content determined by analysis. Based on sludge metal addition, experiments at the Luddington and Lee Valley Experimental Stations indicated that, in general, copper was slightly less toxic than zinc to red beet, lettuce and celery whereas nickel was about twice as toxic as zinc for red beet and lettuce and nearly five times as toxic for celery (3). Based on the extractable metal content of the soils, zinc was twice as toxic as copper for red beet and four times as toxic for lettuce. Nickel was from 3 to 4 times as toxic as zinc for red beet and $2\frac{1}{2}$ times more toxic for lettuce. For celery, nickel was nearly 14 times as toxic as zinc whereas copper was non-toxic to this crop.

3. SLUDGE MANAGEMENT

Sludge application to agricultural land must take into account its nitrogen and phosphorus content, the needs of the particular crop and the quantities of metals being applied. If crop phytotoxicity problems are to be avoided, a cumulative limit must be employed for zinc, copper and nickel applications. Sewage sludges of domestic origin have relatively low concentrations of most metals although they may contain appreciable quantities of zinc. Where industrial effluents also enter the sewers the resulting sludge may contain a range of other potentially toxic elements which could limit its value for agriculture use. The phytotoxic metal usually present in greatest quantity in sewage sludge is zinc, followed by copper and nickel. As mentioned earlier, the relative toxicities of these metals are known to vary according to soil type and reaction, the crop and the basis of the assessment. Furthermore, evidence that the effects of these three metals are additive are inconclusive and unlikely to be complementary, at least not until the concentrations of the individual metals exceed their toxic thresholds.

Nevertheless, to ensure crop safety and to provide guidance on the quantities of any particular sludge that may be safely applied over a long period of time, for convenience the effects of zinc, copper and nickel are considered to be additive. To cater for the most sensitive of crops it is assumed that copper and nickel are 2 and 8 times respectively as toxic as zinc. Their concentrations, after allowing for the toxicity of copper and nickel relative to zinc, are added together to give a Zinc Equivalent for any particular sludge. Up to 560 kg/ha of this calculated Zinc Equivalent can be added over a long period to arable land maintained at pH 6.5 or above, or to pure grass swards if the pH is maintained at pH 6.0 or above without affecting crop growth

Example: A sludge containing in its dry matter 800 mg/kg of zinc,
200 mg/kg copper and 25 mg/kg nickel will have a Zinc Equivalent of:
$$800 + (2 \times 200) + (8 \times 25) = 1400 \text{ mg/kg}$$

Thus 1 tonne of this sludge dry matter contains 1.40 kg Zinc Equivalent. A
total amount of 400 tonnes/ha of sludge dry matter (the permitted
560 kg/ha Zinc Equivalent divided by 1.40, the Zinc Equivalent per tonne of
sludge) can safely be added over a long period (30 years or more). Up to
one-fifth of this safe quantity of sludge may be applied in a single dressing. If one-fifth of the total permitted is applied in any one year, no
more should be applied for a further 5 years. It is necessary to keep
records of the quantities of sludge and the toxic metal contents of each
batch applied to every field so that the total quantities of metals applied
over a period of years can be calculated. However, sludge cannot
accurately be applied to land and metal contents can be variable. It is
therefore essential to check sludge metal additions at intervals by monitoring the increase in total metal concentrations in the soil. If the analysis
indicates greater metal additions than those expected from the quantities
of sludge applied based on Zinc Equivalent, any further additions should be
adjusted accordingly.

4.   SOIL METAL CONCENTRATIONS

For sludge treated soils a very good correlation has been found to exist
between the "total" metal content determined by digestion with nitric/
perchloric acid and the EDTA extractable metal content for zinc, copper
and nickel. The linear regressions of total (y) against EDTA extractable
(x) metal contents are given in the following equations:-

Zinc   y = 1.028x + 141.28 $R^2 = 0.952$

       If x = 130, y = 275

Copper y = 1.245x + 21.71 $R^2 = 0.963$

       If x = 70, y = 109

Nickel y = 3.227x +22.63 $R^2 = 0.857$
       If x = 20, y = 87

Recommended safe maximum concentrations of zinc, copper and nickel in soils

deduced from the toxic thresholds for EDTA extractable metal content are
shown below:-

	Zn	Cu	Ni
	(mg/litre)		
Total metal content (by HNO_3 or $HNO_3/HClO_4$ digested	275	110	85
Extractable metal content (by extraction with 0.05 M EDTA)	130	70	20

A good correlation between total and extractable trace elements in sludge
treated soils was found by Davis (2). It might be expected therefore that
there would be little difference between the relationships of these two
parameters with concentrations in plant tissue. Results confirmed this
trend and in general, the correlation between metals in soil and metals in
plant tissue was not very predictable. This must be very largely due to
the fact that conventional extractants are insensitive to those variable
soil properties such as pH, CEC and organic matter which are known to be
important in determining trace element uptake by crop plants.

5.   MONITORING OF METAL ADDITIONS

In sludge treated soils, additions of metal must be checked at intervals
by monitoring the increase in total metal concentrations since this can
be the only absolute and constant measure of metals present. In the situation where sewage sludge is being applied at regular intervals over a long
period, the extractable metal content will be variable and may not reflect
the true situation in the long term. The only reliable assessment of the
level of metal contamination in soils in the long term is their total metal
content determined by strong acid digestion. However, if there are metals
present which are not of sludge origin, such as from minespoil, it would
be desirable to determine the EDTA extractable to ensure that they do not
exceed the recommended safe maxima.

The total metal addition may be determined by subtracting the normal background concentration from the measured total soil metal concentration.
Normal background total zinc, copper and nickel contents of uncontaminated
agricultural soils are shown as follows:

Zinc	Copper	Nickel
	(mg/litre)	
70	15	25

	or (kg/ha)*	
140	30	50	$(mg/l \times 2$ based on 2×10^6 litres in 20 cm cultivated horizon).

In the example quoted on page 5, the sludge contains 0.8 kg Zn, 0.2 kg Cu and 0.025 kg Ni per tonne of dry matter. If the total safe quantity of sludge (400 tonne/ha) calculated from the Zinc Equivalent is applied, the soil will have received 320 kg Zn, 80 kg Cu and 10 kg Ni per hectare equivalent to 160 40 and 5 mg/litre of Zn, Cu and Ni respectively. When added to the normal background concentration these additions would give theoretical total metal concentrations of 230 mg/litre Zn, 55 mg/litre Cu and 30 mg/litre Ni which are within the recommended safe margins given in the previous section.

Metal accumulations will require to be monitored at intervals depending on the frequency of sludge application, normally every 3 or 4 years, or longer if sludge is applied at more infrequent intervals than this. It is suggested that soil samples are taken for analysis just prior to an intended application of sludge which allows the maximum equilibration time for metals from the previous application.

Metals of mining origin are reported to be more inert than those of sewage sludge origin and in soils contaminated with minespoil, the ration of "total" to EDTA extractable Zn, Cu or Ni will be much wider than in sludge treated soils and as such will be less available to plants. In some instances, although not normally recommended, sludge may be applied to minespoil contaminated land which makes the interpretation of safe total maximum soil concentrations difficult. In these situations it is necessary to determine the EDTA extractable zinc, copper and nickel contents of the soils to ensure that they are well below the safe toxic thresholds before sludge is applied.

Note: The method of monitoring the accumulation of sludge-borne metals which is described and suggested safe maxima for zinc, copper and nickel are not official MAFF recommendations. The system is being put forward for consideration by interested parties.

TABLE I. Crop growth related to extractable zinc levels in soil

	Extractable Zn (mg/kg) in soil *	
Crop	Poor Growth	Satisfactory growth
Chrysanthemum	93	37
Wallflower	154	56
Sugar beet	96	84
Bedding plants	150	100

*Extracted with 0.5 M acetic acid.

TABLE II. Addition of zinc and copper and extractable Zn and Cu contents of plots.

Total Zn or Cu addition (1948-50) (mg/kg soil)	Extractable metal concentration (mg/kg soil) * (1961)	
	Zinc	Copper
280	90	60
360	150	80
800	300	250

*Extracted with 0.5 M acetic acid.

TABLE III. Yields of red beet dry matter when sewage sludge containing copper was mixed with soil.

Copper conc'n (mg/kg soil)	Dry matter yields		
	Leaves and Stems	Roots	Total
	Grammes dry matter per pot		
0	11.3	35.7	47.0
80	8.2	9.4	17.6
160	5.2	3.5	8.7
240	1.0	0.3	1.3

6. REFERENCES

1. Cunningham J D, Keeney D R and Ryan J A. 1975. Yield and metal composition of corn and rye grown on sewage sludge amended soil. J. Environ. Qual. 4 : 448-454.

2. Davis R D. 1979. Uptake of copper, nickel and zinc by crops growing in contaminated soils. J. Sci. Food Agric., 30, 937-47

3. Marks M J, Williams J H and Chumbley C G. 1977. Field experiments testing the effects of metal contaminated sewage sludges on some vegetable crops. MAFF Reference Book 326, "Inorganic pollution and agriculture", pp 235-251 (HMSO, London).

4. Mitchell G A, Bingham F T and Page A L. 1978. Yield and metal composition of lettuce and wheat grown on soils amended with sewage sludge enriched with cadmium, copper, nickel and zinc. J. Environ. Qual. 7 : 165-171.

SLUDGE APPLICATION TO LAND - OVERVIEW OF THE CADMIUM PROBLEM

Jens Aage Hansen & Jens Chr. Tjell
Department of Sanitary Engineering
Building 115,
The Technical University of Denmark
DK-2800 Lyngby, Denmark

Summary

A summary report is given on several activities on the effects of sludge in agriculture in Denmark, especially on the behaviour of cadmium in the soil-plant system. Cadmium is a mobile trace metal in the terrestrial environment. The source for cadmium in soils is probably of minor importance in determining the plant uptake of the metal, compared to the governing parameters pH, soil type and presence of competing cations. Plants receive a substantial amount of cadmium with aerial deposition.

Sludge is a minor source for cadmium in Danish agriculture, while fertilizers and aerial deposition account for more than 95% of the total input.

Generally, the cadmium content of Danish top soil is slowly increasing in time, leading to an increasing human intake of the toxic metal.

If normal Danish sludges (3-5 ppm Cd in TS) are utilized as phosphorous fertilizers, the input to land of cadmium from fertilization will approximate the input from normal inorganic fertilization.

---

* This paper is originally chapter 9 in:
  Sludge Application to land. I: Overview.
  Jens Aage Hansen & Jens Chr. Tjell, Polyteknisk Forlag. Lyngby, Denmark, published in 1982.
  The Danish version of the same publication is available from the publisher: Slammets Jordbrugsanvendelse I. OVERBLIK. Polyteknisk Forlag, Lyngby, Denmark, 1981.

I. CADMIUM

I.1 General

Cadmium is a toxic trace metal, which follow zink closely in
nature, as the chemistry of the two elements are quite alike.
Divalent cadmium ions Cd^{2+} do not participate in precipitation
reactions or form stable chelates under normal, natural conditions. Cadmium is low in the lyotropic row for binding of metals to soil components, and thus this metal is rather mobile
in the soil/plant system.

Danish agricultural soils contain cadmium in the range 0.03-
0.9 ppm, with a median concentration of 0.22 ppm, TJELL & HOV-
MAND 1978. Generally it is assumed that cadmium, like zink, is
mainly sorbed on clay and humus in an exchangeable form in competition with other cations, H_3O^+ and Ca^{2+}. Sludges contain typically 3-15 ppm of cadmium in dry matter with a median of 7 ppm,
(cfr. chapter 2). Much higher concentration have been found in
Denmark (up to 70 ppm) and other countries (over 1000 ppm) as a
result of specific industrial discharges. Agricultural utilization of sludge thus always enhances the concentration of the
element in the soil. An application of 100 t dry matter with 7ppm cadmium
will in average double the concentration in the ploughing layer.

As a result of a large global consumption of zink, the unavoid-
able side product cadmium is consumed in relatively large quantities, mainly in pigments (CdS, CdSe) and stabilizers for PVC-
plastics. The applications are assumed to have caused a tenfold
increase in the environmental turnover of the element, NRIAGU
1979. The principal sources for environmental contamination are
mining and refining of zink ore, solid waste incineration, power
production from coal and use of phosphatic fertilizers.

I.2 Biological significance

The increased presence of cadmium in the general environment is
probably the most serious pollution problem caused by emissions
of inorganic compounds. This statement is prompted by the fact

that measured alimentary intakes of cadmium in many industrialized
countries are approaching the critical limit. The main effect of
too high intakes is an irreversible accumulation of the element
in the kidney tissue, leading to partial or complete malfunction
of the organ. Except for locations with high industrial emission,
only mammals with long lifespans seem vulnerable to this condition,
especially man seems exposed.

I.3 Cadmium in the environment

The relatively high load of cadmium to the environment has also
been demonstrated in danish investigations on the turnover of the
element in agricultural soils. Figure   1 summarizes the results
for in- and outputs of the element to an average danish agricultural soil, as it is reported in FOKUSERING 4.

The largest inputs to soil are with fertilizer (phosphatic) and
through atmospheric precipitation. The outputs are much smaller,
only loss with drainage being significant. Losses via sale of
produce are hardly noticeable. The balance indicate that soil concentrations of cadmium in Denmark are steadily raising, in average
0.6%/year.

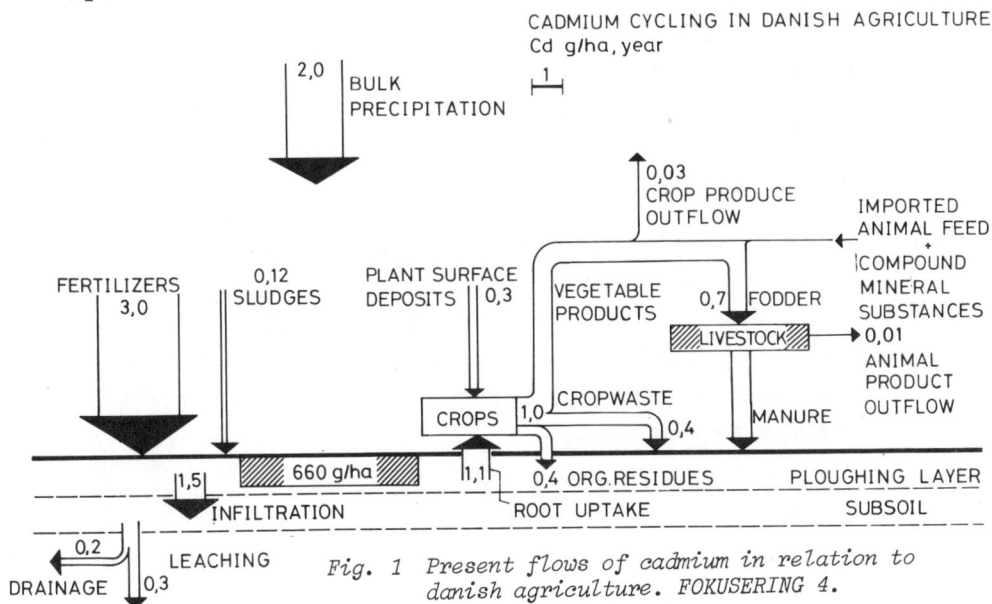

Fig. 1 Present flows of cadmium in relation to
danish agriculture. FOKUSERING 4.

Sewage sludge utilized in agriculture only supply limited amounts
of cadmium to soils in average. Fields actually receiving sludge
may on the other hand receive rather significant amounts. This is
exemplified in table   1, counting the local situation after application of 1 or 10 t/ha dry sludge, compared to the average situation.

I.4 Transport of cadmium through soil

Although cadmium is a rather mobile element in soil (relative
to other trace metals), it is still strongly bound to soil
particles, mainly as adsorbed or ion exchanged. Whatever the
dominating state, it is characteristic that an increase in total concentration in the soil result in an increase in concentration in the soil solute, as it is shown by CHRISTENSEN 1980.
For two danish subsoils (B-horizons) the pH dependant connection
between soil and water concentrations were determined, as shown
in figure   2. Based on these results a model was set up, simu-
lating the most likely transport rate and concentrations of cadmium in soils. Even under unfavourable conditions, following a
large input of cadmium to an acid, sandy soil, the soil water
concentrations at 1 m's depth would not exceed 10 µg Cd/ℓ (figure   3), and further that acceptable concentrations for drinking
water would not be exceeded in 2 m's depth.

It may be concluded, that high loadings of cadmium to soil do not
lead to excessive contamination of groundwater.

Another feature of the rather low transport rate of cadmium in
soil is the correspondingly high retention time for the element
in soil layers. The average retention time for cadmium in the
ploughing layer may be 200-800 years. Consequently contaminated
soil stay high in concentration for extended periods, only very
slow leaching may be expected.

Cadmium in soil
mg/kg TS

Soil solution, µg Cd/l

– – – – loamy sand (Hornum)

———— sandy loam (Ødum)

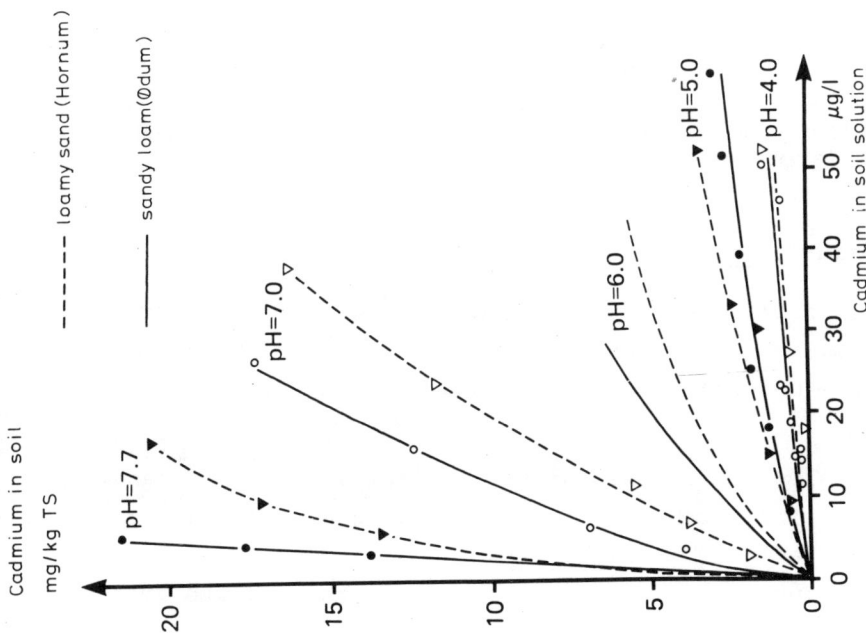

FIG. 2 Relation between soil- and soil solute concentrations of cadmium for two typical danish subsoils at different pH. [Ca] = 10⁻³ M/l. CHRISTENSEN 1980

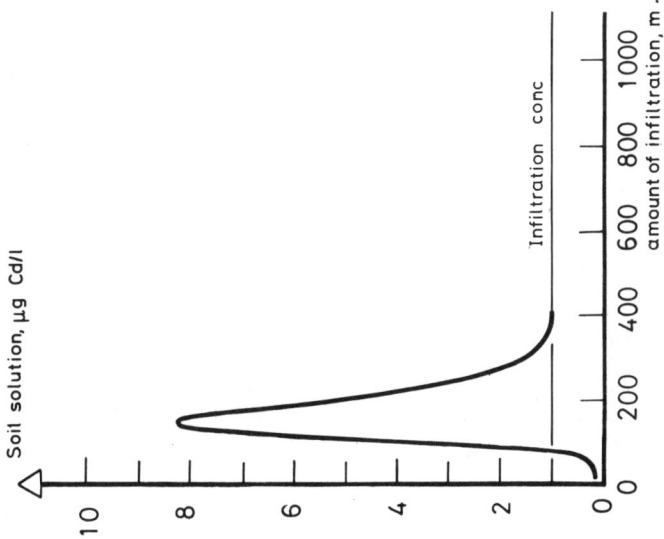

FIG. 3 Modelling of cadmium concentrations in soil solute. Concentrations at 1 m's depth under a ploughing layer added 10 kg .Cd/ha (= +3ppm). pH = 5.5. [Ca] = 10⁻³ M/l. DELRAPORT 9.

Table 1. Inventory for the future cadmium balance on a typical
           danish agricultural area, receiving type-sludge, 1 or
           10 t DM/ha,y.

	Whole Denmark 1) t/y	1 t/ha,year 2) g/ha,y	10 t/ha,year 3) g/ha,y
Inputs:			
Chem. fertilizer	9	0	0
Concentrates	0.3	0	0
Atmosphere	6.9	2.3	2.3
Sludge	0.8	7	70
	17.0	9.3	72
Outputs:			
Produce	0.1	0.7	0.7
Infiltration	4.5	1.5	1.5
	4.6	2.2	2.2
Soil accumulation	12.4	7.1	70
rel. increase 4)	0.6%/y	1.1%/y	11%/y

Notes: 1) Future situation for $3 \cdot 10^6$ ha receiving fertilizer and
           manure.

        2) Sludge is utilized as a phosphorous fertilizer,
           20 kg/ha,y.

        3) Sludge is utilized as nitrogen fertilizer, disregarding
           cumulative effects. No fertilizer or manure used.

        4) Cadmium in soil 0.22 ppm ≈ 660 g/ha in the ploughing
           layer.

I.5 Cadmium in fertilizer and sludge

Phosphatic fertilizers contain cadmium present in the raw phosphate used, sedimentary phosphates being highly variable in content, while magmatic phosphates usually are virtually free of
cadmium.

Cadmium in low content sewage sludge (domestic type) presumably
originate mainly from corrosion of zink in plumbing and gutters,
where especially older zink types contained considerable amounts
of cadmium (up to 1%). Diffuse sources of this nature are difficult to control, but slow replacements in other materials (PVC),

or purer zink qualities, will eventually lower the cadmium content of sludges. The public awareness of cadmium as a potentially dangerous element has further reduced industrial discharges, voluntarily or reluctantly. Combined efforts to reduce cadmium emissions to the sewer have resulted in significantly lower concentrations in sewage sludge. In 1981 most domestic type sludges may contain 3-4 ppm in dry matter, and 10 ppm ought never to be exceeded, FOKUSERING 5C

In table 2 are shown concentrations of cadmium in various potential fertilizers, including sludge, manure and chemical fertilizers based on various raw phosphates. The cadmium content is referred to the phosphorous content of the material.

Table 2. Cadmium concentrations in different fertilizing materials, including sludge. Concentrations shown for the dry matter and in relation to the phosphorous content.

Product	Cadmium in sludge raw phosphate ppm in DM	Cadmium in phosphorous ppm	Note
Sludge	7	350	1
	3-15	150-750	2
Manure	(~0.3)	30	3
Fertilizer based on raw phosphates from:			4
USA (Idaho)	40-340	250-2000	
North-Africa (Marocco)	7-60	40-350	
USA (Florida)	14	85	
USSR (Kola)	0.1-1	0.5-6	
Danish phosphatic fertilizer (1979)	16-22	110-165	5

Notes: 1) Typical figures
2) Probable range for sludges used in agriculture.
3) Based on figures in FOKUSERING 4.
4) Based on raw phosphate analyses, USEPA 1978, assuming 15% P, and a loss of cadmium in manufacture of 10%.
5) Informations from SUPERFOS 1979. Fertilizers marketed in Denmark are mainly on raw phosphates from USA (Florida) and North-Africa.

The range of concentrations of cadmium in phosphorous in chemical
fertilizers is very wide. The very contaminated american types have
only limited local use, while the danish supply of fertilizers is
based on raw phosphates with a medium content of cadmium.

Type-sludge contains 7 ppm cadmium in dry matter, which is around
10 times higher than manure, and around 2.5 times higher than chemical fertilizer, if calculated on phosphorous basis. If sludgephospho-
rous were to contain cadmium as an average chemical fertilizer, the
average content in sludge dry matter should be around 3 ppm. This
low figure is already quite common, especially at small sewage works
as reported by SKOV 1981 and PETTERSSON & ERICSSON 1979.

If type-sludge is to be used as a phosphorous fertilizer, (approx.
1 t DM/ha,year, table 1, the addition of cadmium with sludge will
be 4 g/ha, year, higher than with chemical fertilizer.

This leads to an average increase in the soil cadmium concentrations
of around 1.1%/year in sludged fields as compared to 0.6% in normall:
fertilized fields.

I.6 Plant uptake of cadmium from soil

The plant uptake of cadmium from soil has been the subject of
numerous investigations in Denmark and elsewhere. The experiences
are however not yet conclusive as to the behaviour of cadmium in
the soil-plant system. The most important factors determining the
cadmium concentration in a plant seem to be:

o Plant species, and time of harvest
o Soil type, texture and pH
o Soil concentration of cadmium
o Possible reduction of availability of cadmium in the soil
o Atmospheric contribution to the plant (particles adhering
  to the plant surfaces).

Differences in the cadmium uptake due to genetic variations are
not easily explainable, but they are quite substantial as table
  3 shows. For the selection of typical danish crops grown under
identical conditions, the highest content in lettuce is 500X the

lowest in barley grain. Generally, starch containing fruits
(grains) are very low in cadmium, while proteinaceous leafy
plants are high.

The influence of pH and soil concentration of cadmium may be illustrated in figure 4, where the uptake in rye grass was measured
following application of sludge. The uptake of cadmium appears
roughly to be proportional to the soil concentration at fixed pH,
while an increase in pH of 0.5 unit lowers the uptake 20-40%. In
relation to figure 2 it is quite conceivable, that the plant
uptake of cadmium is strongly dependant on the solubility of the
element in the soil. The results in figure 4 are similar to
results obtained by many foreign investigators, e.g. WILLIAMS
1977, although there are many conflicting informations, as discussed
in depth by CHANEY et al. 1977, and CAST 1981.

The influence of the soil concentration on the plant uptake of
cadmium may also be seen in figure 5, together with the signi-

Table 3 Concentrations of cadmium in danish crops, grown
under identical conditions. DELRAPPORT 3.

Crop	Cadmium concentration ppm in DM
Barley, grain	0.012
Barley, straw	0.14
Oat, grain	0.025
Oat, straw	0.18
Potato	0.32
Kale	0.34
White cabbage	0.22
Carrot, root	0.35
Pea	0.043
Pea, pod	0.082
Pea, straw	0.24
Green bean, pod	0.040
Green bean, straw	0.40
Lettuce	5.2
Spinach	3.9

ficance of the harvest time. The repetition of the experiment
over a total of three years further shows that the availability
of cadmium does not alter within this span of time after sludge
application.

The influence of soil type is shown in figure   6. At the two
employed pH it is obvious that the plants take up the more cadmium at the same cadmium concentration, the lighter the soil.

The plant availability of cadmium may possibly change over time.
The most common assumption is that it decreases strongly over
the years, for instance after a sludge application. This is most
clearly demonstrated by HINESLY et al. 1979, and is also one of
the conclusions of a very extensive recent litterature survey by
DAVIS & COKER 1980. A Swedish experiment employing sludge with
radioactive cadmium shows however that this conclusion may be too
hasty, LÖNSJÖ 1980. Except for the first year of eight consecutive
growth seasons, the plant uptake of cadmium from the sludge in a

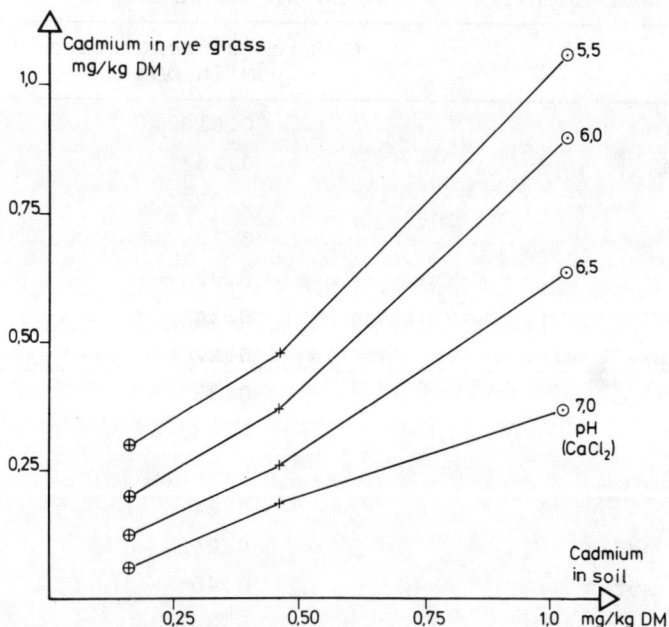

Fig. 4 Cadmium uptake in rye grass from soil with added sludge. Pot
experiment, with pH control. Askov soil. DELRAPPORT 3D.

single application was constant in the whole period. This conclusion is valid only if all parameters as pH were kept constant. A liming was shown to lower the uptake of cadmium, although the sludge cadmium retained its availability relatively to the indigenous cadmium in the soil. This Swedish experiment is in our mind the most convincing evidence showing cadmium in soil only to be bound reversibly, without significant reversion of the availability over at least eight years after a sludge application. For other sources of cadmium to soil, as phosphatic fertilizer on atmospheric dust, there are yet no direct evidence for a similar behavior of the element.

Turning to danish field experiments with cadmium in sludge, the effects of a single application of sludge were followed for six years.

The experimental conditions are described by DAMGAARD-LARSEN et al. 1979a. The results for cadmium in three crops are shown in figure 7 - 9. Although the scatter in resulting cadmium concentrations

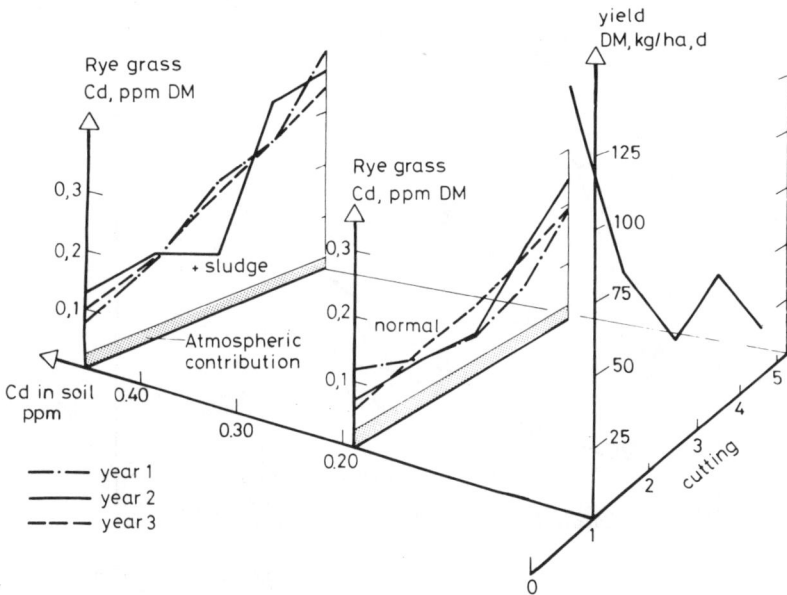

Fig. 5 *Cadmium uptake in rye grass (5 cuts/year) for 3 consequtive years after one application of sludge. Frame experiment with Askov soil. DELRAPPORT 3E. Probable atmospheric contribution is shown shaded.*

in the plants is large, these field experiments support the conclusions arrived at in the pot- and frame experiments (figures   4 - 6).
The inconsistencies in the field experiments are mostly due to
variations in soil pH. For instance liming was carried out at Askov
before the fourth season (5 t lime/ha). Further, the analytical error
for cadmium determination in low content plants (ås barley grain,
figure 7) is quite significant. On top of these problems, the
atmospheric contamination of green plants is of considerable and
varying size over the years. In the figures in this chapter the
estimated limits of this source for cadmium in plants are shown in
shaded areas. (c.f. section I.7)

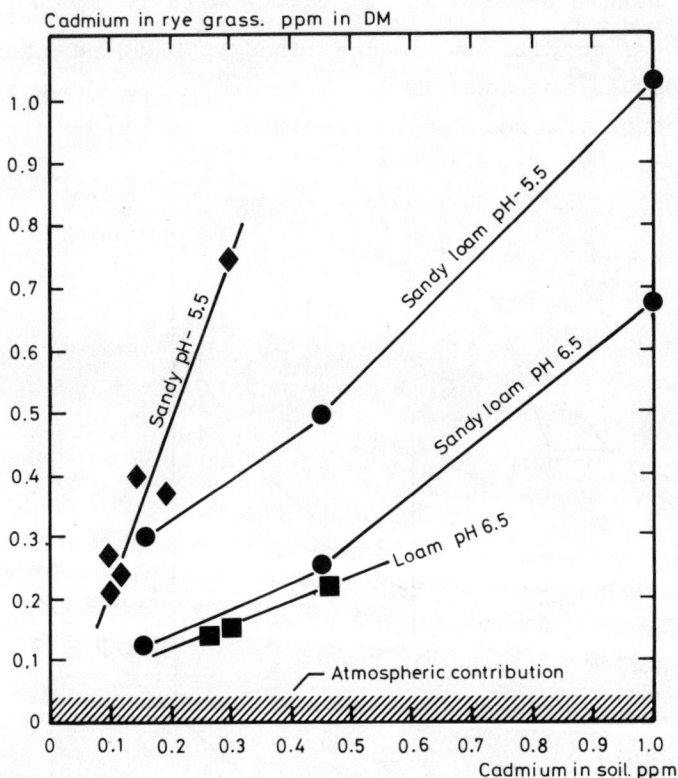

*Fig. 6   Compilation of data on uptake of cadmium in rye grass on 3 soil
types added sludge. Results from pot- and frame experiments. pH was
controlled. DELRAPPORT 3D and 3E. Probable atmospheric contribution is
shown shaded.*

The few examples from danish pot- frame- and field experiments
shown in this chapter seem representatively to illustrate the
most important characteristics of the plant uptake of cadmium.
The genetic factor is the most influential, followed by pH, where
an increase leads to a diminished uptake. As a main rule it can
be assumed that the plant uptake of cadmium is roughly proportional
to the soil cadmium concentration for otherwise identical conditions, and that the plant availability does not change significantly with time.

I.7   Plant uptake of cadmium from the atmosphere

Green plants may receive cadmium directly from the atmosphere
through surface contamination. This has been examined by HOVMAND
et al. 1982 for a few selected crops grown in a background area.
The results are presented in table   4, as the absolute and
percentagewise atmospheric contribution to the cadmium concentration in the plant material. These results are used in the figures in this chapter as shaded areas indicating the upper and
lower limit of the likely atmospheric contribution.

Table   4.   Plant uptakes of cadmium from the atmosphere. Re-
             sults from experiments on isotope dilution of cad-
             mium in soils. HOVMAND et al. 1982.

Plant material	Total cadmium ppm in DM	Atmospheric cadmium	
		ppm in DM	% of total
Rye grass	0.10-0.11	0.04-0.05	30-55
Barley, grain	0.08	0.04	41-58
Barley, straw	0.31	0.08	13-38
Wheat, grain	0.16	0.035	~21
Wheat, straw	0.47	0.14	~30
Rye, grain	0.09	0.02	18-29
Rye, straw	0.51	0.09	11-23
Carrot, root	0.25	0.12	37-59
Carrot, leaves	0.79	0.37	36-58
Cabbage, leaves	0.11-0.16	0.06-0.08	36-60
Cabbage, stem	0.16-0.18	0.07-0.09	28-61

This atmospheric contamination of crops with cadmium is an additional complication in assessing properly the relation between soil concentrations and plant concentrations in normal growth experiments involving moderate amounts of sludge and cadmium. If the atmospheric contribution of cadmium in the figures 7 - 9 is correctly assessed, then even the field experiments lend support to the hypothesis of proportionality between plant and soil concentrations of this element. If the atmospheric contributions to plants are high or varying, even more uncertain relations may be found.

I.8 Cadmium and the terrestrial biota

Microorganisms in soils are not very sensitive to increased soil concentrations of cadmium if pH is not too low (<5). At lower pH enzymatic activities seem to be depressed, e.g. phosphatase activity at pH = 3 and 1-5 ppm of cadmium in a forest floor, TYLER & WESTMAN 1978.

Larger organisms in soil appear to accumulate cadmium. ANDERSEN 1979 and DELRAPPORT 8 have shown that earth worms may contain 100% more cadmium in the tissues (dry weight) than the soil they live in. Despite this, the worms appeared not to be affected, and they were more numerous in soils receiving sludge. For unknown reasons the worms in the sludged area had lower cadmium concentrations than worms from normal soils.

Larger herbivores are likely to take up cadmium in proportion to the feed concentrations, and the target organs for this element, liver and kidney, accumulates cadmium nearly in proportion to the intake, cf. TJELL et al. 1981 and CEC & GSF 1981. Combined with short lifespans, the resulting organ concentrations are low, and normally the animals are not considered affected. For normal conditions, even the consumers of meat are not likely to receive larger amounts of cadmium, MILJØMINISTERIET 1980. This conclusion is also valid for areas receiving large doses of cadmium in sludge, e.g. in USA, CHANEY 1981, although grossly contaminated kidneys may pose a special problems.

—Fig. 7 Uptake of cadmium in barley grain, from 2 soil types, followed in the field 6 years after one application of 30 t/ha dry matter of sludge (L = Lundgård, sandy. A = Askov, loam. Indices 1 or 2 are for sludge with low or high cadmium concentration). The soil concentrations are measured; they should have been nearly doubled according to the calculated supply. Shown with permission from State Research Station Askov. The experiment is described by DAMGAARD-LARSEN et al. 1979a.

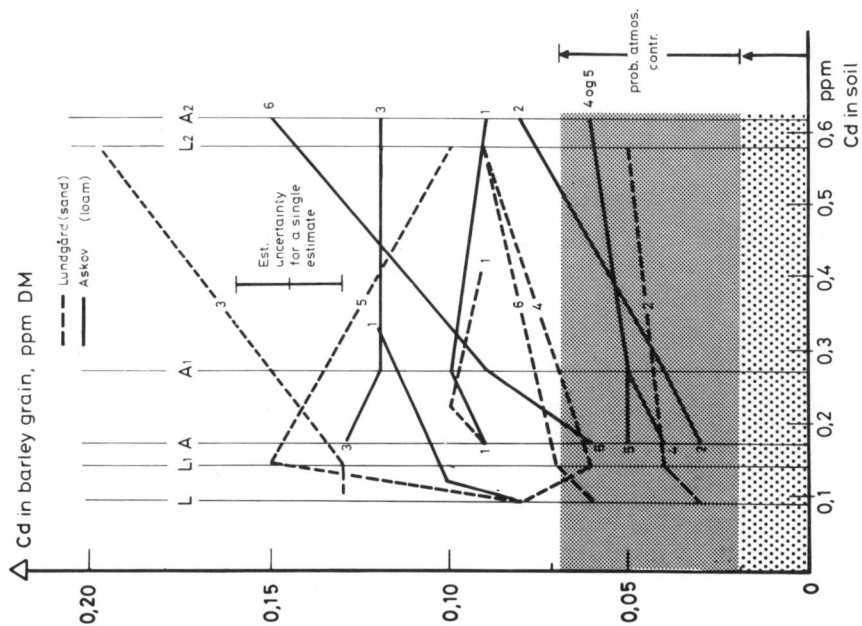

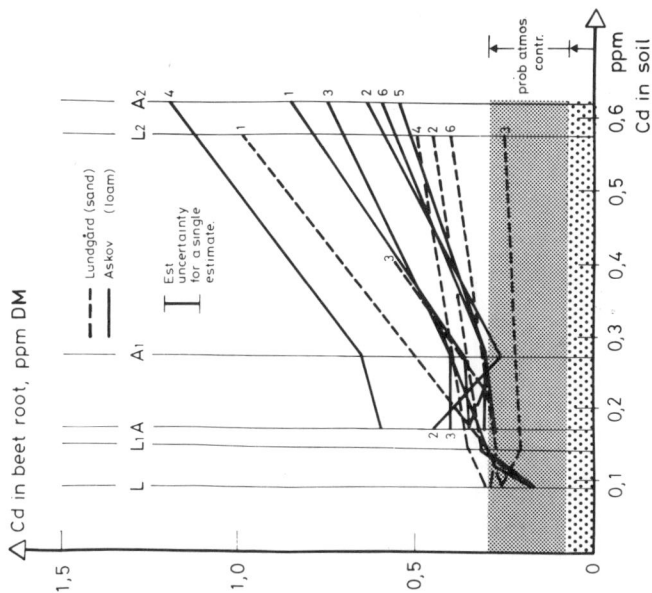

Fig. 8 Same as fig. 7, but for rye grass

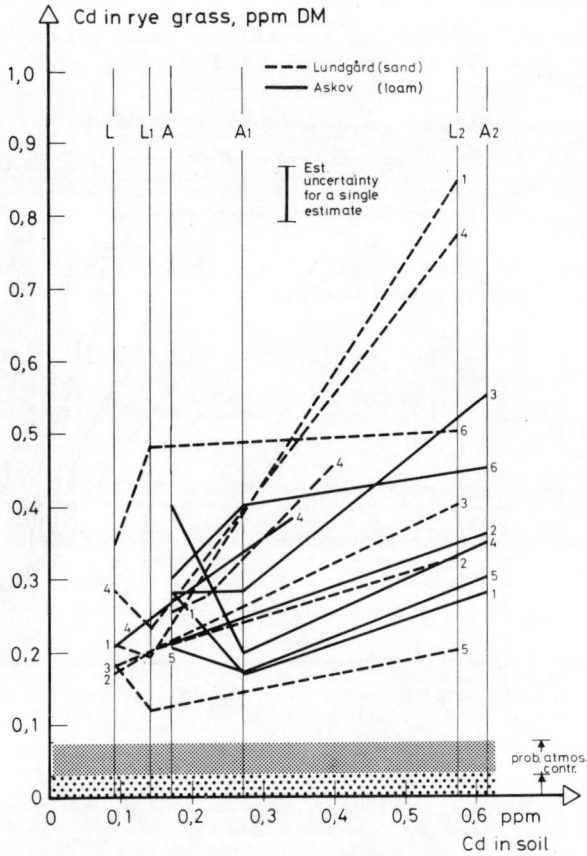

Fig. 9 *Same as fig. 7, but for beet root.*

I.9 Human intake of cadmium

The human intake with food and through breathing is a rather well
studied subject. The use of sludge in agriculture may lead to a
future increase in the average intake of cadmium from food for the
population in general. But sludge is only one source of cadmium in
agriculture, along with fertilizers and atmospheric dust and rain.

The present average human intake of cadmium with food in Denmark
is shown in table 5. Most of the food cadmium is supplied through
vegetabilia. The average of 30 µg/P,d most likely corresponds to
a very wide range of individual intakes. Compared to the provisional tolerable weekly intake of 4-500 µg/P ($\approx$65 µg/P,d), FAO/WHO
1978, the present cadmium intakes in Denmark may be regarded as

Table 5. Alimentary intakes of cadmium for Danes, average
           and calculated for vegetarians.

	Total intake µg/P,d	Distribution %		
		Vegetable	Animal	Water
Normal	<32	64	30	<5
Vegetarian	<33	83	13	<5
Excretion	25			

nearly alarmingly high. The limit value may be too low, as argued by
e.g. CHANEY 1981, or too high as put forward by LAUWERYS 1981. Any
conclusion to this discussion has not been reached yet.

As cadmium is a mobile trace metal all through from soil to human
food, it is possible to predict the future intake of the element
via food for the danish population in average. The prediction in
figure   10 is made as if the balance in figure   1 is to continue
for the next century, FOKUSERING 5C. From the prediction it is assumed that the average human intake of cadmium may increase from
30 µg/P,d to around 50 µg/P,d over 100 years, or 70% higher.
The main part of this increase stems from the accumulation of
cadmium due to continued use of phosphatic fertilizers and atmospheric precipitation, while continued agricultural utilization of sludge may be responsible for approx. 4% of the increase
(= 1 µg/P,d). It is then assumed that type sludge is applied
quite extensively throughout the agricultural area.

If average human food was to originate exclusively from areas
receiving sludge, 10 t DM/ha,y, the average food intake would increase to 58 µg/P,d after 10 years, an increase of 90%. This hy-
pothetic calculation shows that even a relatively low application of sludge may enhance the food intake of cadmium to the limit
set up by FAO/WHO, if a person were to be fed from sludged fields.

These predictions on the human food intakes of cadmium imply that
the consumption and quality of phosphatic fertilizers continue un-

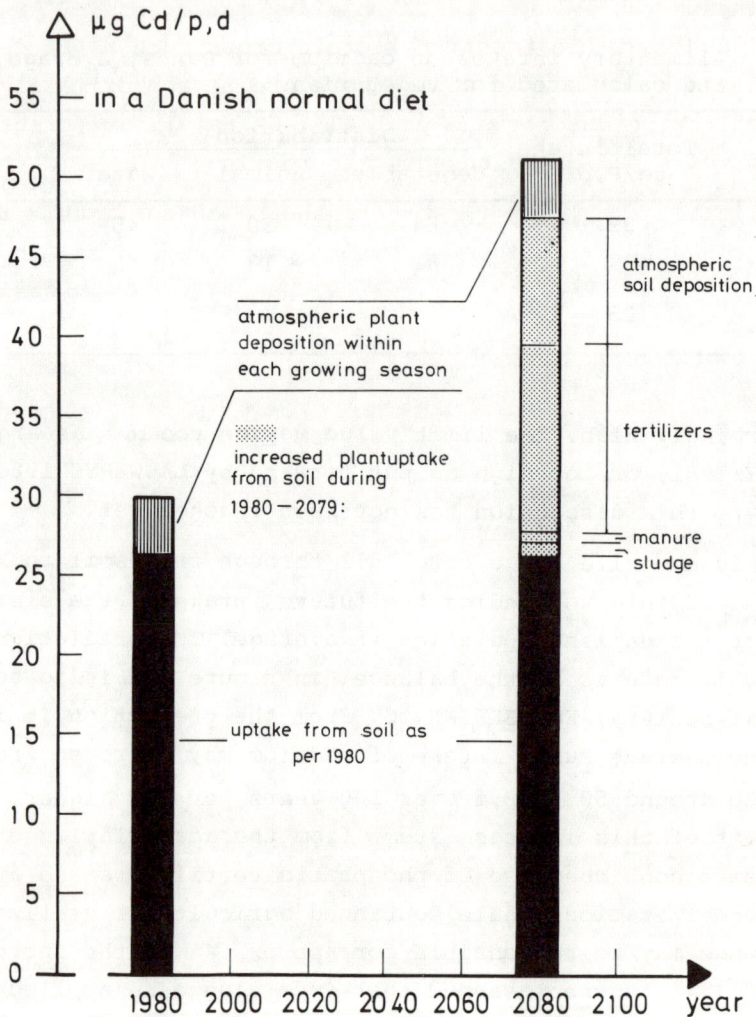

μg Cd/p,d

55 — in a Danish normal diet

Fig. 10 Prediction of the future Cadmium content in the average Danish diet, assuming continued inflow of Cadmium to agricultural soils at present level. FOKUSERING 5C.

changed. It may be foreseen that the phosphorous application rate may be lowered in future agricultural practices, and thus lead to lower increase rate for the food intake of cadmium.

CONCLUSIONS REGARDING CADMIUM IN SLUDGE

Sludge utilization - as opposed to e.g. disposal by landfilling - would be recommendable mainly due to the high phosphorus value.

Typically phosphorus amounts to 2% on a dry matter basis for conventional, secondary treatment. For grain crops such as barley the necessary dressing of sludge would therefore be approximately 1 ton dry matter per hectar and year (1 t DM/ha,y). For practice 5 t every 5 years may prove an adequate routine.

Sludge utilization would on the other hand be objectionable mainly due to its content of cadmium which upon plant uptake may accumulate in food chains. Presently a typical Danish household waste water sludge holds 7 ppm of cadmium; but indications are that 3 ppm is achievable and already manifest in a number of municipal sludges.

Based on these two major considerations table 6 has been established to indicate cadmium balances for one hectar of land supplied in three alternative ways with 20 kg P/ha,y. This amount of phosphorus is to compensate for the typical average removal of 15-20 kg P/ha,y for grain crops in Scandinavia.

Table 6 indicates that use of sludge as well as chemical fertilizers is likely to cause a certain annual increase of soil cadmium. Manure, however, could imply no increase of soil cadmium, if applied only according to phosphorus demand.

In any case the soil cadmium increase seems controllable, if phosphorus be applied according to minimum plant supply. Where sludge and chemical fertilizers are used, their cadmium concentrations should be lowered to the extent possible. For fertilizers this means a purer raw material and/or new production methodology. For sludge it becomes necessary to limit the use of cadmium wherever possible in households and industry.

Application of sludge or manure only to extent that phosphorus is sufficiently supplied implies a deficiency in nitrogen and potassium that must be compensated through chemical fertilization. Such procedure seems logical, because phosphorus is in limited supply and has a much higher purchase price than both nitrogen and potassium. Also, the above cadmium consideration would support the lowest possible use of phosphorus.

Table 6   Cadmium budget for 1 hectar of agricultural land sup-
          plied with phosphorus in three different ways. Unit
          applied is grammes of Cd/ha,y.

Cadmium source		20 kg P in sludge	20 kg P fertilizer	20 kg P manure
PRESENT	Atmosphere	+2.3	+2.3	+2.3
	Fertilization	+7	+3	+0.7
	Crop	-1.5	-1.5	-1.5
	Wash out[1]	-0.7	-0.7	-0.7
	Accumulation	7.1	3.1	0.8
FUTURE	Atmosphere[2]	+1	+1	+1
	Fertilization[3]	+3	+3	+0.7
	Crop	-1.5	-1.5	-1.5
	Wash out[1]	-0.7	-0.7	-0.7
	Accumulation	1.8	1.8	-0.5

Notes 1) Wash-out from soil is highly dependent on soil type and pH; an
          average is indicated.
       2) Future atmospheric fall-out is expected to decrease compared
          with to-day due to increased emission control; arbitrary guess.
       3) Lower cadmium concentrations, e.g. 3 ppm on a dry matter basis,
          seem feasible based on empirical evidence.

Pathogens in sludge may impose certain demands on treatment,
storage, handling application time, tilling of land, choice of
crops and time of harvest or grazing by cattle. Technically these
challenges are not particularly difficult to cope with, but to the
management expenses may be relatively high.

In conclusion, sludge utilization instead of sludge disposal has
its price, both technically and economically. Utilization seems
tied to sludge phosphorus, and it is likely that time and development will favour increased sludge phosphorus utilization. The major reasons for such prediction are that use of cadmium will decrease; tertiary waste water treatment will increase and hence
amounts of sludge phosphorus, possibly at relatively lower prices;
and market prices for chemical fertilizers will continue to rise.

REFERENCES, LITERATURE

ANDERSEN, C. 1979: Cadmium, Lead and Calcium Content, Number and Biomass in Earthworms (*Lumbricidae*) from Sewage Sludge Treated Soil. Pedobiologia, 19, 309-319.

CAST 1981: Effects of sewage sludge on the cadmium and zinc content of crops. Council for Agricultural Science and Technology - U.S. Environmental Protection Agency. Cincinnati, Ohio, USA. (CAST Report No. 83) (EPA-600/8-81-003).

CEC & GSF 1981: Commission of the European Communities & Gesellschaft für Strahlen- und Umweltforschung mbh: Ecotoxicology of Cadmium. Brussels, Luxembourg. München-Neuherberg, BRD. (EUR 7499 EN) (GSF-Bericht Ö-629).

CHANEY, R. L. et al. 1977: Plant Accumulation of Heavy Metals and Phytotoxicity Resulting from Utilization of Sewage Sludge and Sludge Composts on Cropland. In: National Conference on Composting of Municipal Residues and Sludges Products, Silver Spring, August 23-25, p. 86-97. Information Transfer, Inc., Rockville, MD, USA, 1978.

CHANEY, R. L. 1981: Health risks associated with toxic metals in municipal sludge. Oak Ridge National Laboratory. Life Science Symposium Series, Gatlingburg, Tennessee, October 4-8.

CHRISTENSEN, T. H. 1980: Cadmium Sorption onto Two Mineral Soils. Department of Sanitary Engineering, Technical University of Denmark, Lyngby. (Rep. 80-1).

DAMGAARD-LARSEN, S.; KLAUSEN, P.S.; LARSEN, K.E. 1979: Engangstilførsel af slam fra rensningsanlæg til landbrugsjord. (Once for all application of sewage sludge to agricultural land). (In Danish with English summary). Beretning 1467 fra Statens Planteavlsforsøg. Tidsskrift for Planteavl, 83, 387-403.

DAVIS, R. D.; COKER, E. G. 1980: Cadmium in Agriculture, with Special Reference to the Utilization of Sewage Sludge on Land. Water Research Centre, Technical Report WRC TR 139, Stevenage, U.K.

DELRAPPORTER. Reference to Institutional Reports (in Danish with English summaries). Vol. III of "Slammets Jordbrugsanvendelse" (Sludge application to land). Polyteknisk Forlag, Lyngby, Denmark.   (published in 1982).

FAO/WHO 1978: List of maximum levels recommended for contaminants by the Joint FAO/WHO Codex Alimentarius Commission. Third Series. Food & Agriculture Organization of the United Nations and World Health Organization, Rome, Italy. (CAC/FAL 4-1978).

FOKUSERING 1-5. Reference to chapters (in Danish). Vol. II of "Slammets Jordbrugsanvendelse". (Sludge application to land). Polyteknisk Forlag, Lyngby, Denmark, 1981.

HINESLY, T.D.; ZIEGLER, E.L.; BARRETT, G.L. 1979: Residual Effects of Irrigating Corn with Digested Sewage Sludge. Journal of Environmental Quality, 8, 35-38.

HOVMAND, M.F.; MOSBÆK, H.; TJELL, J.C. 1982: Plant uptake of airborne cadmium. (Accepted for publication in Environmental Pollution).

LAUWERYS, R. 1981: Toxicology: General Overview. European Toxicology Forum, April 6-9. International Conference Center, Geneva, Switzerland. (Preliminary issue p. 366-372). Toxicology Forum, Inc.

LÖNSJÖ, H. 1980: Kadmiums växttillgänglighed under fältförhållanden. Några resultat från fleråriga försök med radioaktive Cd-isotoper. (The plant availability of cadmium under field conditions. Some results from field trials of several years duration with radioactive isotopes, in Swedish with English Summary). In: Kadmium i vår odlingsmiljö. (Cadmium in the Soil-Plant Environment). The Royal Swedish Academy of Agriculture and Forestry, Stockholm, Sweden. (Report No. 4).

MILJØMINISTERIET 1980: Cadmiumforurening (Cadmium Pollution, in Danish with English Summary). National Agency of Environmental Pollution, Copenhagen, Denmark.

NRIAGU, J.O. 1979: Global Inventory of Natural and Anthropogenic Emissions of Trace Metals to the Atmosphere. Nature, 279, 409-11.

PETTERSSON, O.; ERICSSON, J. 1979: Tungmetaller och avloppsslam i jordbruket (Heavy metals and sewage sludge in agriculture, in Swedish). Aktuelt från Lantbruksuniversitet, 274. Uppsala, Sweden.

SUPERFOS 1979: Personal communication by P. Eichner, Superfos, Vedbæk, Denmark.

TJELL, J.C.; CHRISTENSEN, T.H.; BRO-RASMUSSEN, F. 1981: Cadmium in Soil and Terrestrial Biota, with Emphasis on the Danish Situation. Internationa Symposium: Cadmium, Interpretation of Data and Evaluation of Current Knowledge, May 6-8, Neuherberg, Federal Republic of Germany.

TJELL, J.C.; HOVMAND, M.F. 1978: Metal Concentrations in Danish Arable Soils. Acta Agriculturæ Scandinavica, 28, 81-89.

TYLER, G.; WESTMAN, L. 1978: Effekter av tungmetallförorening på nedbrytningsprocesser i skogsmark VI. Metaller och svavelsyra (Effects of heavy metal pollution on decomposition in forest soils. IV Metals and sulfuric acid, in Swedish with short English symmary). National Swedish Environment Protection Board, Solna, Sweden. (Report SNV PM 1203).

USEPA 1978: Cadmium Additions to Agricultural Lands Via Commercial Phosphatic Fertilizers. U.S. Environmental Protection Agency. Washington, D.C., USA. (EPA Report SW-718).

WILLIAMS, C.H. 1977: Trace Metals and Superphosphate: Toxicity Problems. Journal of the Australian Institute of Agricultural Science, 43, 99-109.

THE INTAKE BY MAN OF CADMIUM FROM SLUDGED LAND

Dr. J. C. SHERLOCK

Food Science Division, Ministry of Agriculture, Fisheries and Food
United Kingdom

Summary

Sewage sludge often contains undesirable substances which may transfer from the sludged land to crops grown on the land or to grazing animals. This paper discusses how cadmium may transfer from sludged land to man's food and describes a study made to investigate the impact of cadmium from sewage sludge on the dietary intake of cadmium from vegetables. The results show that people consuming vegetables grown on sludge land containing high concentrations of cadmium had higher than normal cadmium intakes from vegetables. No intakes in excess of the PTWI were observed. The results highlight the dangers of estimating intakes in the absence of reliable data on food consumption.

1. Introduction

Sewage sludge is used to advantage in agriculture to increase the organic matter and nutrient contents of soils and thereby increase the yields of crops. In addition to this benefit there are potential hazards associated with its use in agriculture. Sewage sludge often contains undesirable substances including persistent organic chemicals and heavy metals. Some of these substances can and do transfer from the sludged land to crops grown on the land or are consumed by grazing animals. In particular cadmium is readily taken up and translocated by crops which are consumed by man. The purpose of this paper is to present evidence about how cadmium from sewage sludge may influence the dietary intake of cadmium by man.

2. Contamination of Food by Cadmium from Soil

Cadmium in sewage sludge disposed of to land may contaminate man's food by several routes, including:

1) directly by deposition of soil onto crops;

2) animals may ingest sludge; thus meat, milk and eggs from the animals may be contaminated by cadmium;

3) cadmium may be taken up by fodder crops which are eaten by animals and products from the animals are eaten by man;

and 4) cadmium may be taken up from the soil by food crops and these crops ingested by man.

This paper is primarily about contamination of man's food by routes (1) and (4). Before discussing the effect of uptake of cadmium by crops on man's dietary intake of cadmium, the possible impact of the other two routes of contamination of cadmium intake should be considered.

Much of the concern about the possible health effects of cadmium is based on the link between the accumulation of cadmium in the kidneys and kidney dysfunction. Man and most other animals tend to concentrate absorbed cadmium in their liver and kidneys. Therefore an animal consuming above average amounts of cadmium either from crops or soil will tend to have higher concentrations of cadmium in its liver and kidneys. Evidence for

this is shown in Table 1, which presents information on the cadmium concentrations in the tissue of cattle known to be receiving a diet containing
higher than normal concentrations of cadmium and compares these values with
the concentration of cadmium in normal cattle tissue. The results show that
the exposed animals have higher concentrations of cadmium in their liver and
kidneys but the concentration of cadmium in their muscle tissue seems not to
be affected. For most people in the UK liver and kidney are not important
items in the diet, however, prudence dictates that it would be best if offal
from animals receiving elevated intakes of cadmium is not consumed by man.
Evidence published elsewhere[1] indicates that increased intakes of cadmium
by lactating cattle do not affect the cadmium content of milk, which in the
UK is uniformly low. Eggs produced by fowl ranging on ground highly contaminated with cadmium /up to 350 mg/kg in the soil/ contained on average
0.01 mg/kg of cadmium which is no different from what is normally observed[2].
Thus contamination of food by cadmium from routes (2) and (3) is unlikely to
be of major importance.

Four years ago the opportunity arose for the Ministry of Agriculture,
Fisheries and Food (MAFF) to make a dietary study of people who were likely
to consume vegetables grown on land which had received massive applications
of sewage sludge during the previous decades. The remainder of this paper
describes the study.

3. Background
Seven Market Gardens /land used for intensive horticulture of cash crops
such as lettuce, carrots and spinach/ in a semi-rural area near London were
known to have received massive applications of sewage sludge over many years.
These applications of sewage sludge had, on most of the market gardens,
elevated the soil cadmium concentrations to well in excess of the currently
accepted UK guidelines for maximum cadmium concentrations in sludged soil,
namely 3.5 mg/l. The mean soil cadmium concentrations on each market garden
lay in the range 1.5 - 14.1 mg/kg with a range of individual measurements
over all the market gardens of 0.1 - 26.2 mg/kg. Cadmium is known to be
readily taken up and translocated by many vegetables and it was considered
likely that the Market Gardeners and their families were likely to consume
more of their own produce than anybody else. Consequently MAFF took the
view that the Market Gardeners and their families constituted a 'critical
group' in respect of exposure to cadmium from vegetables. That is they were
considered to be likely to be consuming more cadmium than is usual. There-

TABLE 1

CADMIUM IN TISSUE FROM CATTLE RECEIVING ELEVATED

INTAKES OF CADMIUM COMPARED WITH CADMIUM

IN NORMAL CATTLE TISSUE

| TISSUE | Mean Concentration of Cadmium in tissue (fresh weight basis) mg/kg | |
	Cattle with elevated intakes of cadmium (a)	Normal Cattle
Liver	0.55 (5)[b]	0.14 (37)
Kidney	1.8 (5)	0.29 (32)
Muscle	0.02 (5)	0.03 (72)

a. Feed contained 1 mg/kg (dry weight) of cadmium, normal values are generally much less than this.

b. Number of samples in parenthesis.

TABLE 2

CADMIUM IN CROPS GROWN ON LAND CONTAINING

ELEVATED CONCENTRATIONS OF CADMIUM COMPARED

WITH CADMIUM IN NORMAL CROPS

| Crop | Mean concentration of Cadmium mg/kg in crop (fresh weight basis) | |
	Crops Grown on Land with elevated cadmium concentration (a)	Normal Crops
Lettuce	0.18 (50)[b]	0.06 (17)
Cabbage	0.04 (45)	0.01 (22)
Spinach	0.63 (11)	0.08 (4)
Carrot	0.15 (2)	0.05 (13)
Potato	0.14 (7)	0.03 (20)

a. Soils containing greater than 1 mg/kg of cadmium.

b. Number of samples in parenthesis.

fore it was decided to investigate the dietary intakes of cadmium by this
critical group.

4. The Investigation

The investigation was divided into 4 parts, namely:

1) representative samples of crops and soil were taken from the
   Market Gardens for analysis for a number of metals including
   cadmium;

2) the Market Gardeners and their families (the participants in the
   dietary study) were asked to keep a record in a specially prepared diary of the weights and types of vegetables they consumed
   during one month, (September). They were also asked to indicate
   which vegetables were 'home grown' and which were not;

3) the participants provided exact replicates of the food (excluding
   drink) which they consumed during one week in September;

4) samples of blood and urine were collected from the participants
   for analysis.

5. Results

Analysis of the Vegetables: The cadmium concentration in certain types of
vegetables collected from the Market Gardens are shown in Table 2 and compared with the cadmium concentrations in vegetables grown on uncontaminated
land. The results show that the vegetables grown in the Market Gardens
contain higher concentrations of cadmium than is usual. Many vegetables,
for example peas, beans, tomatoes and courgettes tend not to accumulate
cadmium as evidenced by information published elsewhere[2].

Cadmium in the Diets: Table 3 presents information on the weights of the
diets, the intake of cadmium from the diets and the concentration of cadmium in the diets, and compares them with the National Average values. The
concentrations of cadmium in the diets provided by the participants were
estimated in two ways. The first method was by analysis of the replicate
diets and the second was by means of calculations using information contained in the diary records and the results of the analysis of vegetable
samples taken from the Market Gardens. The estimates of the weekly intake
of cadmium reflect both the concentration of cadmium in the diets and the

weights of food consumed. Because of this there appears to be little difference between the participants' cadmium intake and the intake from the National Average diet. The estimates of mean cadmium concentration in the participants diets show that they were exposed to higher dietary concentrations of cadmium than is usual. The participants' dietary intakes of cadmium were all less than the FAO/WHO Provisional Tolerable Weekly Intake (PTWI) for cadmium of 400 - 500 ug.

6. Discussion

Many of the vegetables grown at the Market Gardens contained elevated concentrations of cadmium as indicated by the information in Table 2 and all of the participants consumed at least some vegetables grown in the Market Gardens. Consequently it is to be expected that there would be a higher concentration of cadmium in the participants' diets than is seen in the average diet and evidence for this has been presented in Table 3. The greater the quantity of vegetables any person consumes the greater will be his or her cadmium intake. If a person consumed 4 kg of normal vegetables each and every day then his or her cadmium intake would be about 400 ug/week, that is the PTWI for cadmium: in the UK the average person consumes about 1.5 kg of food (excluding drink) each day. A high cadmium intake from uncontaminated vegetables is possible in theory but almost impossible to achieve in practice. The participants in this study consumed vegetables containing elevated concentrations of cadmium and the effect of this on their intake of cadmium from 'home grown' vegetables is shown in Figure 1, which also shows how the average person's intake of cadmium varies with his or her consumption of fresh vegetables. The experimental points in the figure show that the more vegetables the participants consumed then the greater their cadmium intake. The straight line fitted to the points indicates that each kilogram of 'home grown' vegetables contributed 64 ug of cadmium to the dietary intake. However, none of the participants consumed more than 3 kg per week of 'home grown' vegetables and this limited their cadmium intakes. This agrees reasonably well with what we have found in other dietary surveys. In fact our studies on food consumption statistics lead us to believe that it would be unlikely for any person to consume more than about 5 kg of fresh vegetables in a week. Even at this very extreme level of consumption people eating vegetables grown on the Market Gardens in this study would barely reach the PTWI for cadmium.

TABLE 3

DIETARY INTAKE OF CADMIUN BY PARTICIPANTS
COMPARED WITH NATIONAL AVERAGE VALUES

Estimate of Dietary intake	Number of Diets	Weight of Solid Food Consumed kg/week	Mean Intake of Cadmium µg/week	Mean Concentration of Cadmium in the diets µg/kg
replicate diet	21	5.0	114	23
diary estimate	21	6.3	180	28.5
National average	16	6.6	112	17

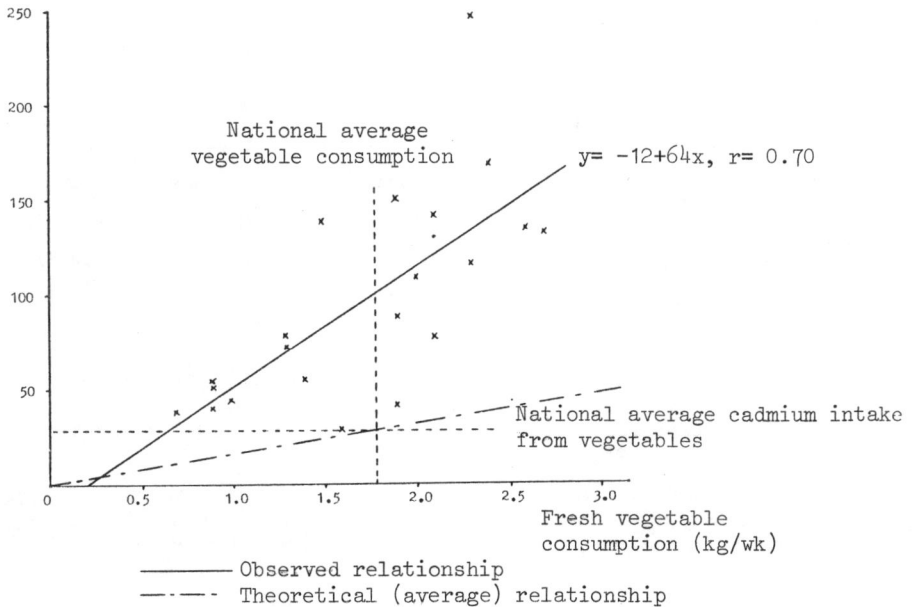

FIG. 1 - Cadmium intake from fresh vegetables - Diary study

The results from this study demonstrate that, before making judgements
about risks to the health of people consuming vegetables containing
elevated concentrations of cadmium or any other contaminant, it is
essential to know the quantity of vegetables the people consume; the same
is true for foods other than vegetables. Making assumptions about the
quantity of vegetables people might consume is all too easy but these
assumptions can be dangerous. Intakes calculated on the basis of ill-
founded assumptions of food consumption may give so much cause for concern
that they lead to precipitate action being taken, when in fact none is
required. Referring back to Table 2 the mean concentration of cadmium in
spinach grown on the Market Gardens was 0.63 kg/kg. If it had been
assumed that people might eat 1 kg of spinach each week, this would have
led to the conclusion that these people would have had a cadmium intake of
at least 630 ug/week, which is in excess of the PTWI (400 – 500 ug). The
results of this simple but misleading calculation might have caused action
to be taken to prevent the sale of vegetables from the Market Gardens. In
the light of the results from the study described in this paper such a
drastic course of action would have been unwise, although the results show
that it would be undesirable to allow the cadmium content of the soil in
the Market Gardens to rise further. The medical examination of the participants showed that none of them were suffering from heavy metal related
disease.

In conclusion, we in the UK adopt the view that the estimation of the intake
of contaminants by 'at risk' individuals must be made in a sound scientific
fashion. To achieve this end it is essential to have reliable information
on the consumption of food by individuals as well as reliable information
on the concentrations of contaminants in food.

REFERENCES

1. Sharma, RP; Street, JC; Verma MP; and Shupe JL : 'Cadmium uptake from
   feed and distribution in livestock products'. Environmental Health
   Perspectives 28, , 59-66, 1979.

2. Report issued by the Department of the Environment, December 1980
   Shipham Survey Committee – Soil contamination at Shipham : Report on
   studies completed in the village and advice to residents.

TOTAL AND BIORELEVANT HEAVY METAL CONTENTS AND THEIR USEFULNESS IN
ESTABLISHING LIMITING VALUES IN SOILS

H. HÄNI and S. GUPTA

Summary

With the help of several glasshouse experiments, using the metals
zinc, copper and cadmium, it is shown that a mild extractant (0.1 M
NaNO$_3$) is best suited to define a uniform value for different soils,
which allows to estimate the metal concentration in a given plant.
The consequences of this finding for the establishment of limiting
values in soils are discussed.

1. Introduction

In many industrialized countries it is nowadays generally accepted
that limiting values for certain substances prove to be efficient
ways and means of protecting soils against pollution. Such limiting
values are mainly significant for substances which are not degraded
and therefore accumulated in soil like the heavy metals. Increasing
amounts of these elements in soil give rise to increased amounts
in plants which - in the case of elements of high zootoxicity like
cadmium - may represent a danger for the consumers.

In Switzerland a new law, namely the law on the protection of the
environment, is discussed in parliament. In this law a chapter
"soil protection" is supposed to be included which constitutes the
basis for an ordinance on limiting values for heavy metals in the
soil. In the following it is tried to show the difficulty in fixing
such values, and some approaches how to overcome these difficulties
are discussed.

2. Material and methods

The pots used in the glasshouse experiments hold 6 - 10 kg of soil.
The soils are characterized by their pH and cation exchange capacity
(CEC). The cadmium experiments were conducted in the framework of

the COST — Action (7). Here it was decided to mix the soil of each pot with 100 g of dry sludge, artificially enriched in cadmium by the addition of cadmium sulfate. In all other experiments the metal addition occured in the form of the corresponding metal salts without any sludge application.

The extraction methods with NH_4OAc (pH 4.8), DTPA (pH 7.3) and NH_4OAc + EDTA (pH 4.65) are described elsewhere (1, 11, 10). For the $NaNO_3$ extraction 20 g of air dried soil were shaken for 2 hours with 50 ml of 0.1 M $NaNO_3$. The metal concentrations in soil extracts and plant material were determined by atomic absorption.

3. Results and discussion

The extent of the heavy metal input into a soil can best be estimated from its total content (9). Therefore, these contents are always used when limitations of the load have to be calculated. However, the soil properties are not taken into consideration by this approach what means that for certain cases, e.g. acid sandy soils, these values may be too high. For this reason, in addition to the measurement of the total content it is proposed to determine a soluble amount which independently of the soil properties can be related to the metal content in a given plant (4, 5, 6, 7, 8).

Taking zinc as an example it is shown in figures 1 — 3 that a mild extractant like 0.1 M $NaNO_3$ is much more successful in separating soils of different metal binding capacities than DTPA (pH 7.3) or NH_4OAc (pH 4.8). The influence of the soil properties on the metal binding is not sufficiently recognized when these latter solvents are used, which is mainly true for the region of low metal addition. Due to complexing and acid properties of these solvents the importance of soil pH is therefore underestimated.

Copper is contrary to zinc preferentially bound to organic matter as — according to figure 4 — the amount which is solubilized by DTPA is much greater than the amount solubilized by NH_4OAc. The ratio of $DTPA/NH_4OAc$ is the higher the richer the soil is in humus. On the other hand, the ratio of $DTPA/NH_4OAc$ is near 1 for zinc.

What is expected from the chemical dissolution experiments is largely confirmed by plant trials. For copper and zinc the amount in soil is calculated where the concentration of the phytotoxic threshold in

the test plant is reached, whereas for cadmium the zootoxic threshold is taken as a critical level because this element is more
toxic for men and animals than for plants. The figures in table 1
show that a content of around 1 ppm zinc soluble in 0.1 M $NaNO_3$
may indicate a critical soil level. When we look at the corresponding
total contents in the three acid soils we see that different limiting
values should be fixed for different soils, whereby it has to be
noticed that the 70 ppm of the acid soil with the lowest cation
exchange capacity are well below the 300 ppm of Kloke. The difficulties occuring with DTPA or NH_4OAc are best recognized by the
fact that 30 ppm in the acid soil give rise to 200 ppm in the plant
whereas 30 ppm in the alkaline soils correspond to much lower plant
values which are furthermore not much higher than the contents in
the control plants (figures in brackets). Similar observations are
made for copper and cadmium (s. table 2 and 3). However, copper
shows a greater variability in the amount soluble in 0.1 M $NaNO_3$
than zinc and cadmium. As already seen for zinc in acid sandy soils
the critical soil levels for cadmium are also distinctly below the
Kloke value of 3 ppm.

It is well known that heavy metals are more easily accumulated in
vegetative than in generative plant parts. Accordingly 0.23 ppm
of $NaNO_3$-soluble cadmium lead to 1 ppm in dwarf bean pods whereas
already 0.025 ppm of soluble cadmium give rise to 1 ppm in dwarf
bean leaves (s. table 4). In the lower part of this table the
corresponding soil concentrations at which 8 ppm (phytotoxic
threshold) of cadmium in the radish foliage are reached are also
calculated.

4. Conclusions

In the situation of a heavy metal excess it seems to be evident
that a mild extractant (e.g. salt solution) is best suited to recognize the soil level which may become critical for a given plant.
Hence, it is proposed to complete the total amount by the amount
of a lightly soluble fraction. Herewith it should be prevented
that limiting values are used which might be too high for unfavourable cases (e.g. acid sandy soils). However, it is indicated
to control the soluble amount also in soils of higher metal binding

capacity because the metal binding ability of a soil may be lowered
due to pH decrease (washing out of lime, acid rainfall) or degradation
of soil organic matter. The soluble soil concentration has finally
to be based on the metal content in edible parts of the most sensitive plants.

De Vries (2) has found that heavy metals are taken up more easily
by plants under glasshouse than under field conditions. Therefore,
the figures obtained in glasshouse experiments can show the difference
between one soil and another; the absolute figures, however, should
be worked out from field experiments.

Literature

(1) Andersson, A.: Relative Efficiency of Nine Different Soil
Extractants, Swedish J. agric. Res. 5, 125-135, 1975

(2) De Vries, M.P.C. and Tiller, K.G.: Sewage Sludge as a Soil
Amendment, with Special Reference to Cd, Cu, Mn, Ni, Pb and
Zn - Comparison of Results from Experiments Conducted Inside
and Outside a Glasshouse, Environ. Pollut. 16, 231-240, 1978

(3) Fleming, G.A.: Not published data within the EEC Concerted
Action, Working Group 5, 1980

(4) Furrer, O.J., Keller, P., Häni, H., Gupta, S.: Schadstoffgrenz-
werte - Entstehung und Notwendigkeit, EAS-Seminar "Landwirt-
schaftliche Verwertung von Abwasserschlämmen, Basel, 24. - 26.
September 1980

(5) Gupta, S., Häni, H. und Schindler, P.: Mobilisierung und Im-
mobilisierung von Metallen durch die organische Bodensubstanz,
Interner Bericht FAC, März 1980

(6) Gupta, S. and Häni, H.: Easily Extractable Cd-content of a Soil -
its Extraction, its Relationship with the Growth and Root Characteristics of Test Plants and its Effect on Some of the Soil
Microbiological Parameters, Proc. of the Sec. European Symp.,
Vienna, October 21-23, 1980 on "Characterization, Treatment
and Use of Sewage Sludge", Published by D. Reidel Publishing
Company, Dordrecht NL, 665-676, 1981

(7) Gupta, S. and Häni, H.: A Preliminary Report on the Standardized Cd-Pot Experiment 1980 (EEC Concerted Action), Meeting on 16 - 17 June 1981 at Portici, Italy

(8) Häni, H. und Gupta, S.: Ein Vergleich verschiedener methodischer Ansätze zur Bestimmung mobiler Schwermetallfraktionen im Boden, Landw. Forschung, SH 37, 267-274, 1980

(9) Kloke, A.: Mitt. VDLUFA, H. 2, März/April 1977

(10) Lakanen, E. and Erviro, R.: A Comparison of Eight Extractants for the Determination of Plant Available Micronutrients in Soils, Acta Agric. Fenn. 123, 223-232, 1971

(11) Lindsay, W.L. and Norvell, W.A.: Development of a DTPA micr-nutrient Soil Test, Agron. Abstr. 69, 84, 1969

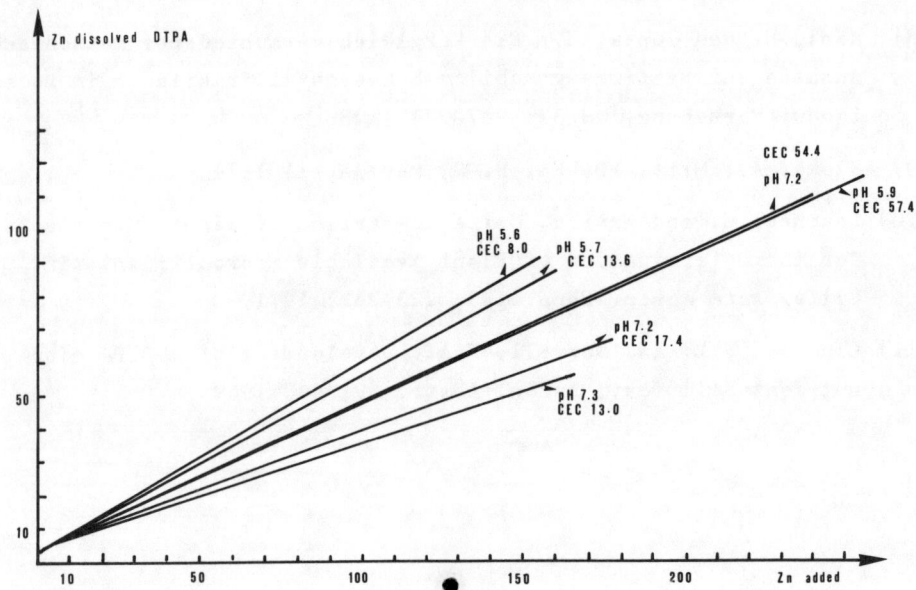

Fig. 1 Zinc dissolved in DTPA as a function of zinc added to different soils

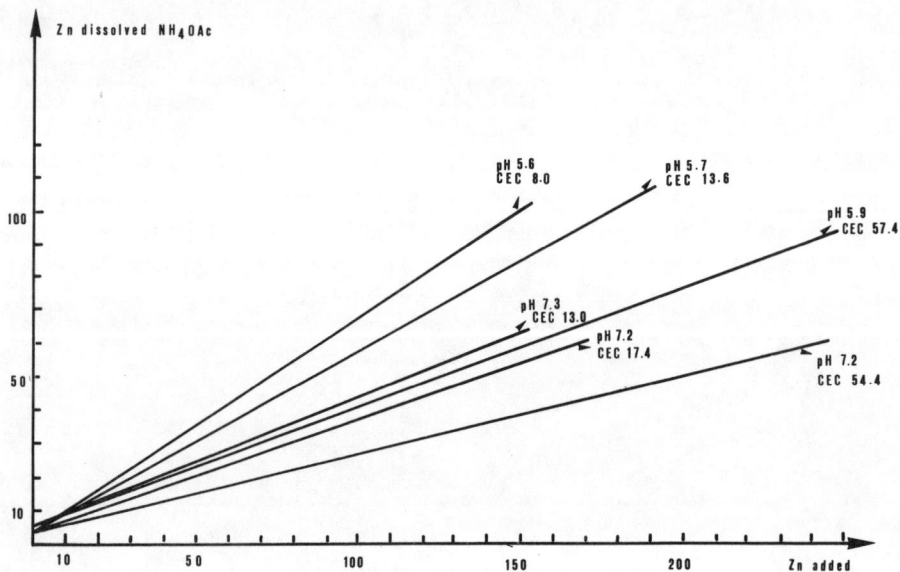

Fig. 2 Zinc dissolved in NH₄OAc as a function of zinc added to different
soils

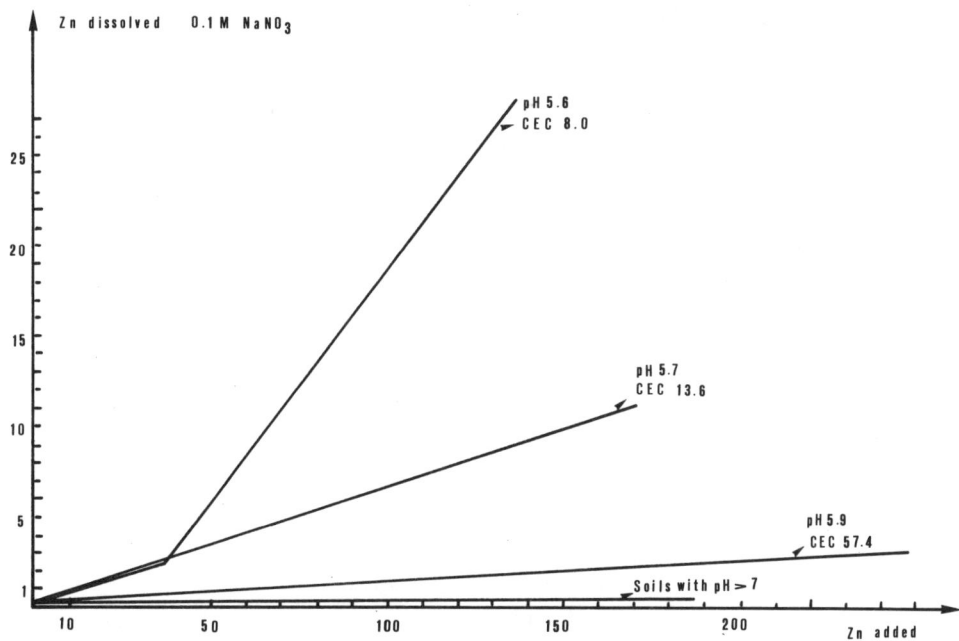

Fig. 3 Zinc dissolved in 0.1 M NaNO$_3$ as a function of zinc added to different soils

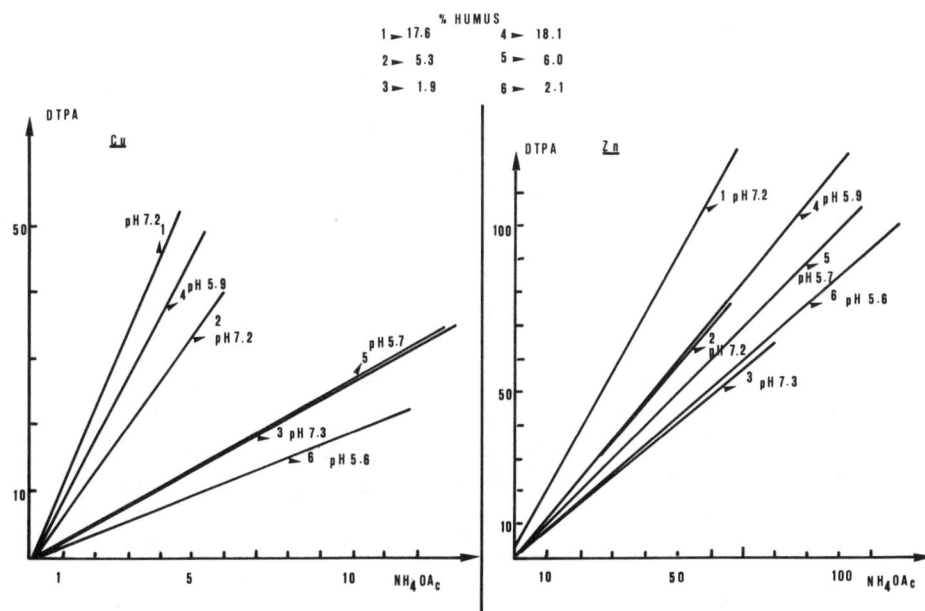

Fig. 4 Comparison of the solubility of zinc and copper in DTPA and NH$_4$OAc

Table 1: The behaviour of red clover in zinc amended soils

| Soil | Zn (ppm) | | | | Zn (ppm) |
	NH_4OAc	DTPA	$NaNO_3$	Total	Red clover
pH 5.9 CEC 57.4	30	30	1.3	218	200 (97)
pH 5.7 CEC 13.6	7	7	0.8	78	200 (132)
pH 5.6 CEC 8.0	8	8	0.8	70	200 (113)
pH 7.2 CEC 54.4	30	55	<0.25	262	96 (78)
pH 7.2 CEC 17.4	30	35	<0.25	130	34 (27)
pH 7.3 CEC 13.0	30	25	<0.25	128	75 (57)

figures in brackets: zinc content in red clover grown on soil
without zinc addition

soil: Kloke value 300 ppm

plant: phytotoxic threshold 200 ppm

Table 2: The behaviour of red clover in copper amended soils

| Soil | Cu (ppm) | | | | Cu (ppm) |
	NH_4OAc	DTPA	$NaNO_3$	Total	Red clover
pH 5.9 CEC 57.4	4.2	38.4	0.66	168	18.5 (15.6)
pH 5.7 CEC 13.6	12.0	32	2.0	127	20 (12.5)
pH 5.6 CEC 8.0	3.5	6.6	1.0	83	20 (12.7)
pH 7.2 CEC 54.4	4.2	50.0	0.20	170	18.0 (8.9)
pH 7.2 CEC 17.4	4.3	29.2	0.20	133	11.6 (10.5)
pH 7.3 CEC 13.0	5.0	15	0.20	102	15.0 (13.2)

figures in brackets: copper concentration in red clover grown on
soil without copper addition

soil: Kloke value 100 ppm

plant: phytotoxic threshold 20 ppm

Table 3: The behaviour of rye grass in cadmium amended soils (results from Fleming, s. Lit. 3)

Soil	Cd (ppm) NH$_4$OAc + EDTA	NaNO$_3$	Total	Cd(ppm) rye grass
pH 5.1 CEC 4.5	0.3	0.06	0.8	1 (0.20)
pH 5.6 CEC 16.6	0.7	0.06	1.2	1 (0.24)
pH 6.9 CEC 15.2	1.2	0.06	4	1 (0.30)

figures in brackets: cadmium concentration in rye grass
grown on soil without cadmium addition

soil: Kloke value 3ppm
plant: zootoxic threshold 0.5 - 1 ppm
phytotoxic threshold 8 ppm

Table 4: The behaviour of different plants (separately specified for generative and vegetative parts) in cadmium amended soils

Soil	Cd (ppm) NH$_4$OAc + EDTA	NaNO$_3$	Total	Cd ppm Dwarf bean pods	Cd (ppm) NH$_4$OAc + EDTA	NaNO$_3$	Total	Cd (ppm) Dwarf bean stalk+fol.
pH 5.7 CEC 10.6	12.8	0.23	12.8	1	1.8	0.025	1.8	1
pH 5.2 CEC 17.1	5.8	0.23	7.5	1	1.0	0.025	1.0	1
				radish roots				radish foliage
pH 5.7 CEC 10.6	0.8	0.01	0.8	1	1.8	0.026	1.8	8
pH 5.2 CEC 17.1	0.4	0.014	0.6	1	1.3	0.05	1.8	8

CADMIUM CONCENTRATIONS IN FIELD AND VEGETABLE CROPS - A RECOMMENDED
MAXIMUM CADMIUM LOADING TO AGRICULTURAL SOILS

M.D. WEBBER AND T.L. MONKS

Wastewater Technology Centre
Environmental Protection Service
Environment Canada
Burlington, Ontario, L7R 4A6, Canada

Summary

Field and vegetable crops were grown in lysimeters on soils treated
with liquid and air-dried sewage sludges. Cadmium loadings to soils varied
with sludge type and application rate but did not exceed 7.4 kg/ha.

Cadmium concentrations in the crops responded similarly to liquid and
air-dried sludge treatments. They appeared to be more closely related to
differences between crop species, plant parts and soil pH than to the
cadmium loadings to soil.

Cadmium concentrations in the crops generally decreased in the order;
Swiss chard and buckwheat, larger than Romaine lettuce, wheat and soybean,
larger than orchard grass. Wheat and soybean grains exhibited lower
concentrations than the respective vegetative materials.

Cadmium concentrations in the crops were highly dependent upon soil pH.
For example, the cadmium concentration in Swiss chard grown on Plainfield
soil (pH 5.6) was 1.42 µg/g for the NPK fertilizer treatment and ≤1.2 µg/g
for the sludge treatments. NPK fertilizer reduced and sludge addition
increased the soil pH.

Evidence is presented that 5 kg Cd/ha added to agricultural soil
represents little if any hazard to the food chain. It is recommended that
this level of addition be considered for international acceptance as the
maximum permissible sludge cadmium loading to soil.

1. INTRODUCTION

The effects of land application of sewage sludge on crop yield and quality and on groundwater quality have been studied using lysimeters at the Wastewater Technology Centre, Burlington. Field crops including orchard grass, wheat, buckwheat and soybean, and vegetable crops including Swiss chard and Romaine lettuce have been grown. One objective of this paper is to examine the effects of different crop, soil and sludge factors on cadmium concentrations in vegetation. A second objective is to make a recommendation to Working Party No. 5 regarding maximum permissible cadmium loadings to agricultural soils.

2. EXPERIMENTAL

Three types of sludge resulting from the addition of alum, ferric chloride and lime to remove phosphorus from wastewaters were applied in liquid and air-dried forms to five soils (Table I) in lysimeter experiments (1). Hereafter, the sludges will be referred to as Al, Fe and Ca sludges, respectively. Cumulative cadmium loadings to soils in the experiments varied with sludge type and application rate but did not exceed 7.4 kg/ha. Cumulative dry matter loadings of the liquid and air-dried sludges ranged up to 390 and 426 tonnes/ha, respectively.

TABLE I. SOIL PROPERTIES

Soil	pH 0.01 M CaCl$_2$	Organic Matter %	CEC* meq/100g
Liquid Sludge Experiment			
Brady fine sand	4.4	1.0	0.6
Caledon loamy sand	7.4	2.2	9.0
Conestoga loam	7.3	4.9	18.2
Air-Dried Sludge Experiment			
Plainfield sand	5.6	2.1	4.7
New Liskeard clay	7.2	6.4	28.2

*Cation exchange capacity

Orchard grass (Dactylis glomerata L.) cv. Frode, Swiss chard (Beta vulgaris var. cicla) cv. Fordhook Giant and Romaine lettuce (Lactuca sativa) cv. Parris Island Cos were grown during the liquid sludge experiment. Spring wheat (Triticum aestivum) cv. Glenlea, buckwheat (Fagopyrum esculentum Moench) cv. Tokyo, soybean (Glycine max (L). Merr.) cv. Harcourt and Swiss chard and Romaine lettuce were grown during the air-dried sludge experiment.

3.   RESULTS

Cadmium concentrations in the crops responded similarly to the application of liquid and air-dried sludges to soils and the following is a resume of results from the two lysimeter experiments.

Crop Effects

Cadmium concentrations in plant materials grown during the lysimeter experiments were related to plant species and plant part. Swiss chard and buckwheat generally exhibited larger concentrations than wheat, soybean and Romaine lettuce as illustrated in Figure 1a. Orchard grass (not shown) generally contained less than 0.3 μg Cd/g. Wheat and soybean grains contained less cadmium than the respective vegetative materials.

Soil Effects

The soils employed in the lysimeter experiments were chosen to represent contrasting textures, pH, organic matter contents and cation exchange capacities (Table I).

Cadmium concentrations in the plant materials were related to soil pH but there was no obvious relationship with the other soil properties. The cadmium concentrations of crops, particularly Swiss chard and Romaine lettuce, grown on the acid Brady soil were much larger than when grown on the neutral Conestoga and Caledon soils (Figure 1b). Similarly, the cadmium concentrations of Swiss chard and Romaine lettuce grown on Plainfield soil treated with commercial (NPK) fertilizer, pH 5.2, were larger than when grown on the same soil treated with sewage sludge, pH 6.3 (Figure 1a and Table II).

It is not known why cadmium concentrations in orchard grass and Swiss chard grown on Brady soil were smaller for the NPK fertilizer treatment than for the sludge treatment (Figure 1b). However, this soil was strongly acid and crops grown on the NPK fertilizer treatment experienced severe stress; Romaine lettuce did not survive and only one replicate of Swiss chard survived. Sludge treatment produced satisfactory growth presumably because one of its effects was to increase soil pH.

Crops grown on the Conestoga and Caledon soils exhibited similar cadmium concentrations (Figure 1b) despite large differences between the organic matter contents and cation exchange capacities of these soils.

TABLE II.   CADMIUM CONCENTRATIONS (μg/g) IN CROPS GROWN ON PLAINFIELD SOIL

Treatment	NPK fertilizer	Fe Sludge Rate		
		low	medium	high
Swiss chard	1.42	1.20	1.44	1.20
Romaine lettuce	1.24	0.83	0.76	0.56
Soil pH (0.01 M CaCl$_2$)	5.2	5.9	6.1	6.3
Cadmium loading (kg/ha)	0	2.5	5.0	7.4

Effect of Sludge Type

Orchard grass and Romaine lettuce grown on Caledon and Conestoga soils exhibited no consistent relationships between cadmium concentration and sludge type as illustrated in (Figure 1c). However, the cadmium concentration of Swiss chard was increased by Al and Fe sludges and reduced by Ca sludge relative to the NPK fertilizer treatment. All three sludges reduced the cadmium concentrations in Swiss chard and Romaine lettuce grown on Plainfield soil compared to the NPK fertilizer treatment and the reductions appeared to be related to increased soil pH (Table III).

TABLE III.   CADMIUM CONCENTRATIONS (μg/g) IN CROPS GROWN ON PLAINFIELD SOIL

Treatment	NPK fertilizer	Sludge		
		Al	Fe	Ca
Swiss chard	1.42	0.78	1.20	0.10
Romaine lettuce	1.24	0.91	0.56	0.39
Soil pH (0.01 M CaCl$_2$)	5.2	6.3	6.3	7.6
Cadmium loading (kg/ha)	0	4.0	7.4	1.5

Effect of Sludge Rate

Sludge application rates to soils in the liquid sludge experiment were were based on N loadings, and repeated applications of 100 (low), 200 (medium) and 300 (high) kg N/ha were applied.

The cadmium concentrations of orchard grass, Romaine lettuce and Swiss chard grown on Caledon and Conestoga soils exhibited little effect of increasing sludge application rates to soil, as illustrated in Figure 1d.

Effect of Discontinuing Sludge Application to Soil

Sludge application to the Caledon and Conestoga soils in the liquid

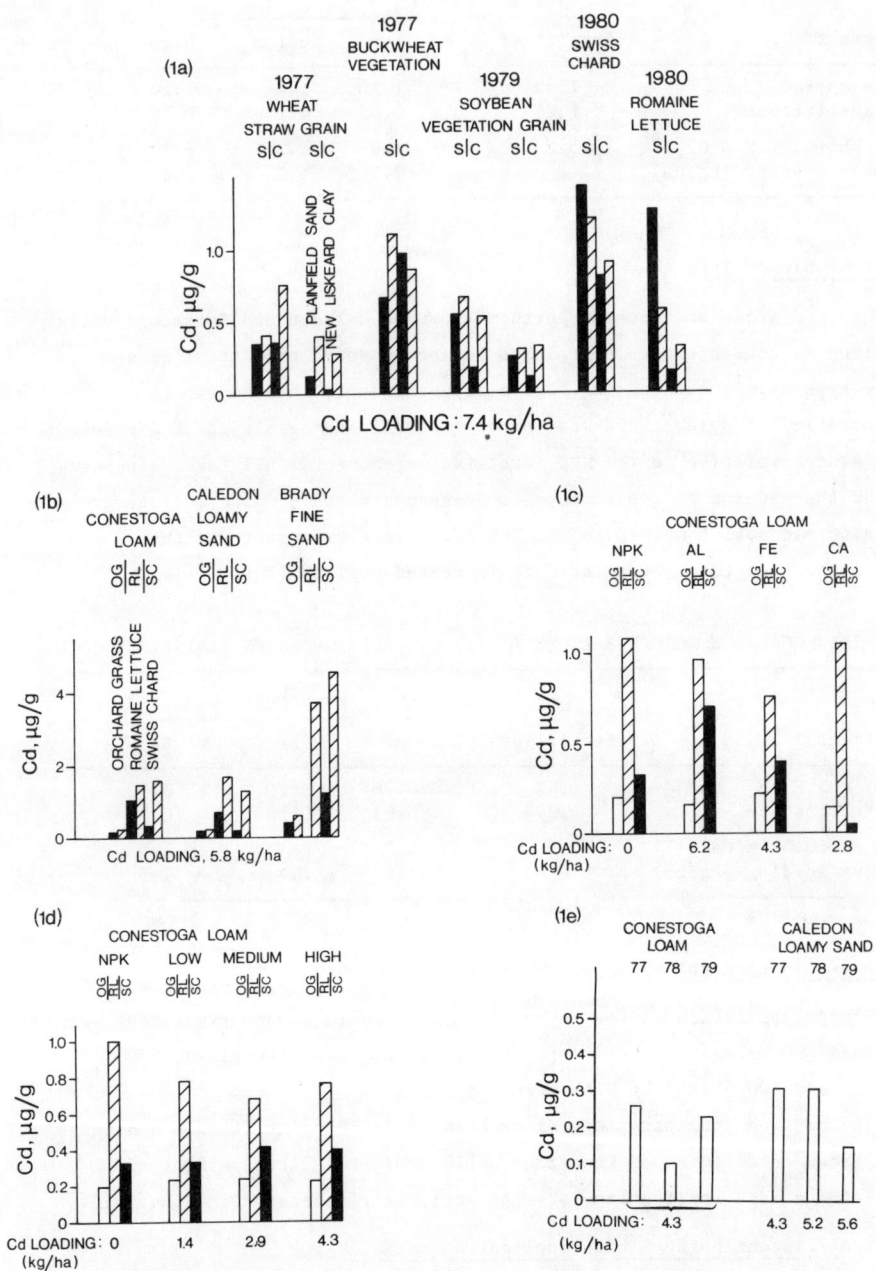

FIGURE 1. Cadmium concentrations in crops grown on sludge treated soils in lysimeters. (In 1a and 1b solid bars are NPK fertilizer treatment and hatched bars are sludge treatment).

sludge experiment was initiated in 1973 and continued until 1977. Following 1977, it was continued on the Caledon soil and was discontinued on the Conestoga soil.

There were no consistent differences between the cadmium concentrations in orchard grass grown on these two soils during 1978 and 1979, as illustrated in Figure 1e.

4. CADMIUM LOADINGS TO SOILS

It is generally agreed that cadmium addition to agricultural land in sewage sludge must be regulated because it represents a hazard to the food chain. However, there is disagreement about acceptable cadmium additions to soils (Table IV). Currently, maximum permissible loadings range from 0.6 kg/ha for Finland to 24 kg/ha for Wisconsin, USA. With the exception of Alberta, Canada and the USEPA, they are independent of soil properties and site conditions.

TABLE IV. MAXIMUM PERMISSIBLE CADMIUM LOADINGS (kg/ha) TO AGRICULTURAL LANDS (2)

Sweden	0.075*	England	5	Ontario, Canada	1.6
Finland	0.6	Switzerland	7.5	USEPA	5-20
Netherlands	1.0	Alberta, Canada 0.8-1.5		Wisconsin, USA	24

* Five-year loading; can be repeated

5. DISCUSSION AND CONCLUSIONS

If nations consumed just their own agricultural produce, cadmium loadings to soils in sewage sludge would be only of national concern. However, international trade occurs on a large scale and it is important that maximum permissible cadmium loadings to soils be standardized to ensure uniform high quality of produce. Such standardization will require international agreement on an acceptable cadmium loading to soil.

Prior to defining permisssible cumulative cadmium loadings in sewage sludge for USA soils, "diet scenario" risk analysis was conducted (3). Based on this analysis, a 5 kg Cd/ha limit was established for acid soils and larger limits were established for soils with pH ≥ 6.5 and cation exchange capacities ≥ 5 meq/100 g. The acid soil risk analysis assumed a model of cadmium exposure to humans from sludged gardens in which; (a) sludge comprised 50% of the dry weight of surface soil, (b) soil pH was 5.5,(c) the individual obtained 50% of his vegetable foodstuffs, including potatoes, from the garden during 50 years and,(d) the individual was a high risk person, with iron, zinc or calcium deficiencies causing higher than normal cadmium absorption. This analysis indicated that high risk individuals have at least a 4-fold protection factor at the 5 kg Cd/ha loading rate on acid soils, and that average individuals have a 12 to 24-fold protection factor. Much larger protection factors would apply where this amount of cadmium was added to soils with pH ≥ 6, as required for land application of sludge in most countries.

A 5 kg/ha limit has also been adopted for cadmium loading in sewage sludge to soils in England and Wales. Following a thorough review of available information, Davis and Coker (4) concluded that if extremes of soil texture and acidity are excluded, then fodder crops, cereal grains, potatoes and the edible parts of leguminous crops present little hazard to the human food chain. Crops whose leafy parts are eaten by humans present the greatest hazard but they comprise only about 0.5% (fresh wt. basis) of the diet for an average United Kingdom consumer.

Experimental results presented in this paper indicated that cadmium concentrations in several field and vegetable crops were more closely related to differences between crop species, plant part and soil pH than to cadmium loadings to soil. Cadmium loadings to the sludge treated soils ranged up to 7.4 kg/ha, however, cadmium concentrations in the crops were frequently larger for the NPK fertilizer treatment which added no cadmium to soil than for the sludge treatments.

Experimental evidence and the USA "diet scenario" risk analysis indicate that 5 kg Cd/ha added to agricultural soil represents little if any hazard to the human food chain. Allowing addition of this large an amount of cadmium to soil recognizes spreading constraints, practical sludge disposal requirements and the fact that disposal on land can be beneficial both agronomically and economically. Thus, it is recommended that 5 kg/ha be considered for international acceptance as the maximum permissible sludge cadmium loading to soil.

REFERENCES
1. Webber, M.D., Y.K. Soon and T.E. Bates, Lysimeter and Field Studies on Land Application of Wastewater Sludges. Water Sci. Tech., 13:905-917 (1981).
2. Commission of the European Communities, Inventory of Rules, Guidelines and Recommendations for the Agricultural Use of Fertilizers and Sewage Sludge. Concerted Action Treatment and Use of Sewage Sludge (COST 68 Bis) (1980).
3. Federal Register, Criteria for Classification of Solid Waste Disposal Facilities and Practices; Final, Interim Final and Proposed Regulations (as Corrected in the Federal Register of September 21, 1979), Vol. 44, No. 179:53438 (September 13, 1979).
4. Davis, R.D and E.G. Coker, Cadmium in Agriculture with Special Reference to the Utilization of Sewage Sludge on Land. Water Research Centre Technical Report, TR 139 (1980).

CADMIUM IN SLUDGE-TREATED SOIL IN RELATION TO POTENTIAL

HUMAN DIETARY INTAKE OF CADMIUM

R.D. DAVIS, J.H. STARK and C.H. CARLTON-SMITH

Environmental Protection Directorate, Water Research Centre,
Elder Way, Stevenage SG1 1TH, Herts UK.

Summary

Results of a field trial growing wheat, potatoes, red beet, lettuce, cabbage and ryegrass have been used to model the relationship between cadmium in sludge-treated soil and potential human dietary intake of cadmium. From this model it is clear that transfer of cadmium from soil into staple crops like wheat and potatoes is much more important than cadmium in such leafy vegetables as lettuce, even though the concentrations found in leafy vegetables may be higher. The model shows that a soil concentration of 6.0-12.0 mg Cd/kg is compatible with the WHO/EPA maximum acceptable dietary intake of cadmium of 70 μg/day for an average consumer taking all his/her crops from soil treated with sludge. There are good grounds for believing that a soil concentration limit in this range is a conservative estimate of what is safe.

1. INTRODUCTION

When sewage sludge is used on agricultural land it is essential that metal contamination problems are averted, preferably by objective limits which ensure adequate environmental protection without causing unnecessary expenditure, either to industry or to authorities responsible for sludge disposal. In the UK, pollution problems are avoided by observance of environmental quality criteria and for sludge utilisation on land these take the form of maximum permissible soil concentrations of metals or the equivalent loading rates of metals in sludge applied to agricultural land[1]. Historically, soil contamination limits have been based on some arbitrary multiple of 'background' concentrations of metals in soil. A sounder and more objective approach would be to base limits on the results of experiments which have defined soil concentrations associated either with phytotoxic effects on crops or with undesirable increases in metal inputs to human or animal diets. For cadmium, it is human dietary intake which is of most concern. This element is a cumulative poison, so subtle increases in the diet sustained over long periods could conceivably lead to toxicity problems in man. Cadmium is also the element which most commonly limits rates of application of sewage sludge to agricultural land. In this paper, a model is described which relates soil concentrations of cadmium to potential human dietary intake of cadmium. Its purpose is to provide an objective concentration limit for cadmium in soils receiving sludge.

2. EXPERIMENTAL

Data used in the dietary model were derived from a field trial at Royston, financed by the UK Department of the Environment, already described by Davis and Stark[2] at the Second European Symposium on Characterization, Treatment and Use of Sewage Sludge organised by the Commission of European Communities in Vienna in October 1980. The present paper draws on results of crop analysis from the third year (1981) of the trial on a calcareous soil of pH value 8.0 and cation exchange capacity 20 meq/100 g. The trial involved the treatments set out in Table I. Liquid sludge (S_1-S_3) plots were established in 1978 and solid sludge (S_4) plots were established in 1979. All treatments together with unsludged control plots were replicated within blocks, there being six blocks on the site making a total of 180 plots in all. This permits six crops to be grown concurrently in a rotation. The crops are winter wheat (<u>Triticum aestivum</u> L. cv. Maris Huntsman), potato (<u>Solanum tuberosum</u> L. cv. Pentland Crown), cabbage (<u>Brassica oleracea</u> L. cv. Stonehead), red beet (<u>Beta</u>

- 138 -

vulgaris L. cv. Crimson Globe), lettuce (Lactuca sativa L. cv. Mildura) and ryegrass (Lolium perenne L. cv. Melle). Each year all the crops are grown (one crop on each block) and samples of soil and plant material are taken for analysis.

Table I. Cadmium added to soil by the sludge treatments (Royston field trial

Sludge type	Treatment	Dry solids addition (t/ha)	Cadmium addition (kg/ha)
S_1	R_1	19	1.3
	R_2	38	2.6
	R_3	76	5.2
	R_4	152	10.4
S_2	R_1	19	0.24
	R_2	38	0.48
	R_3	76	0.95
	R_4	152	1.90
S_3	R_1	18	0.6
	R_2	35	1.2
	R_3	71	2.3
	R_4	141	4.6
S_4	R_4	250	10.5
	R_5	500	21

Sludges S_1-S_3 were applied in 1978 and sludge S_4 in 1979.

3. RESULTS

    Fig. 1 shows the very close relationship between cadmium in sludge originally added to the soil in 1978/79 and cadmium concentrations (extracted by concentrated nitric acid) persisting in soil in 1981. Soil pH values on sludge-treated plots were in the range 7.1-8.2. Fig. 2 shows cadmium concentrations in crops in 1981. In all cases there was a close linear relationship between cadmium concentrations in soils and crops. Highest cadmium concentrations were seen in lettuce and lowest concentrations occurred in potato tubers. The other crops formed an intermediate group with cadmium concentrations in the range 0.05-0.70 mg/kg dry matter.

Fig 1. Metal concentrations in soil vs. metal addition to land 1981

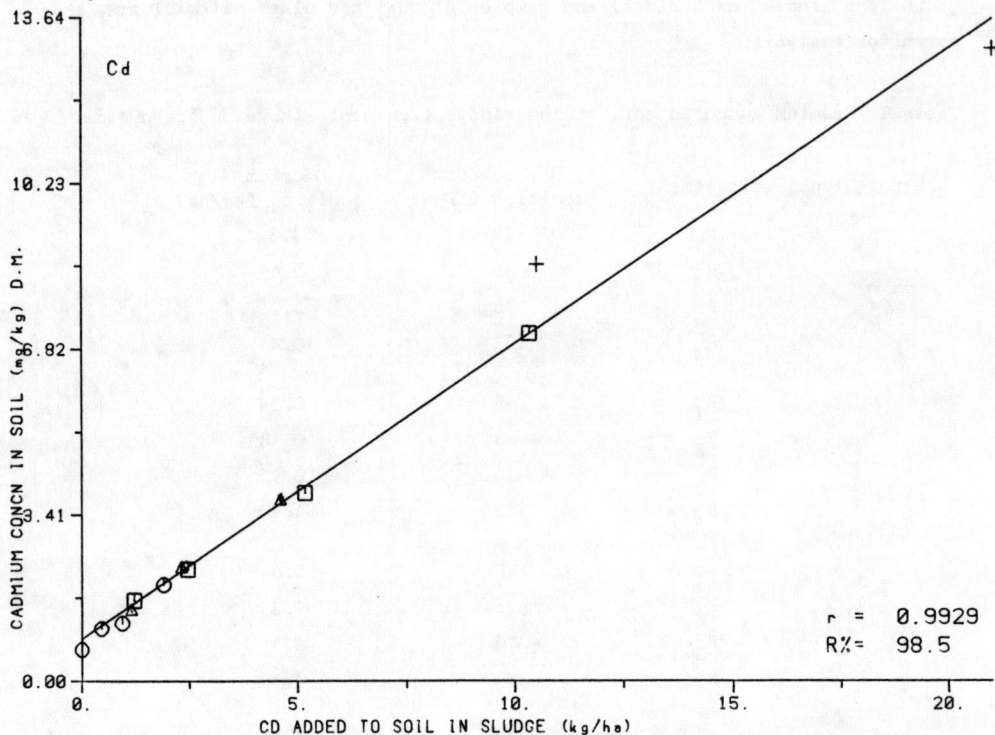

Cd

CADMIUM CONCN IN SOIL (mg/kg) D.M.

13.64

10.23

6.82

3.41

0.00

0.   5.   10.   15.   20.

CD ADDED TO SOIL IN SLUDGE (kg/ha)

r = 0.9929
R%= 98.5

KEY to Figs 1 and 2

symbol   sludge type

S1-Perry Oaks
S2-Hogsmill Valley
S3-S1/S2 mixed
S4-solid P.O.

r   product-moment
      correlation coefficient

R%   percent variance accounted
      for by the regression

Fig 2. Cadmium concentration in crops and soil 1981

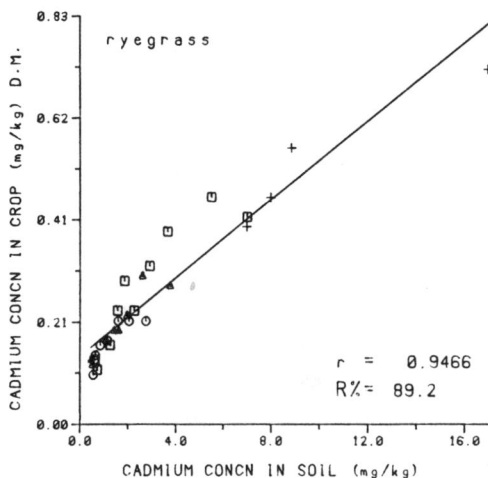

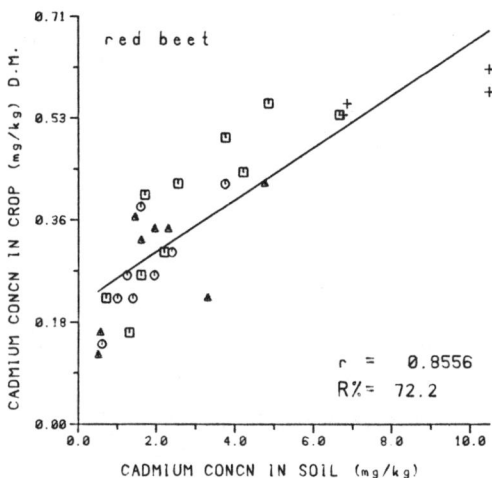

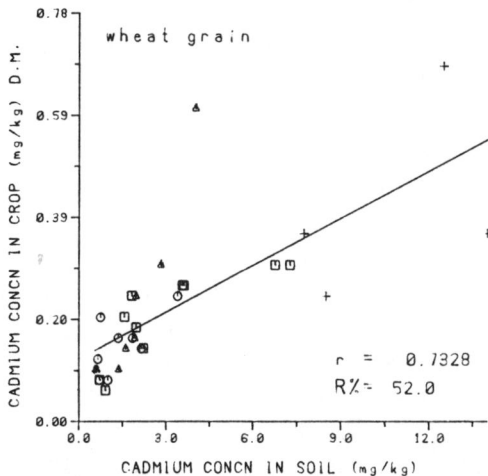

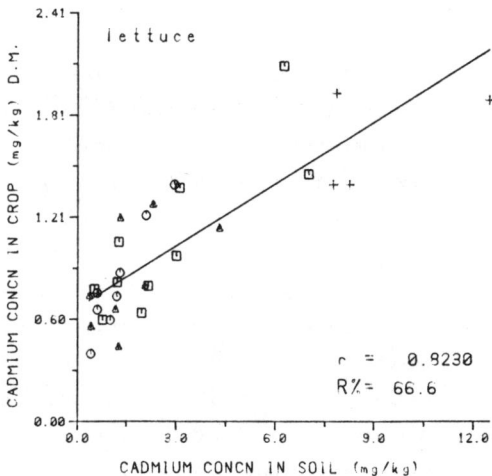

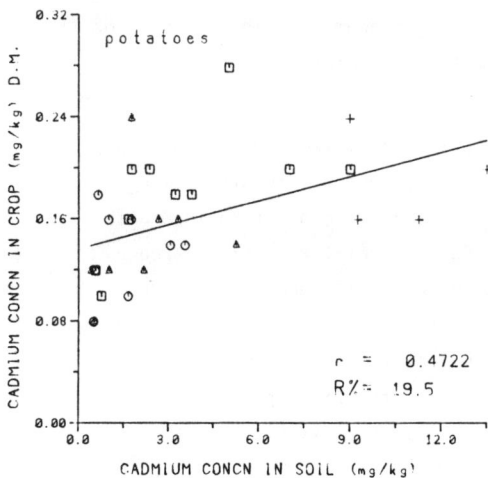

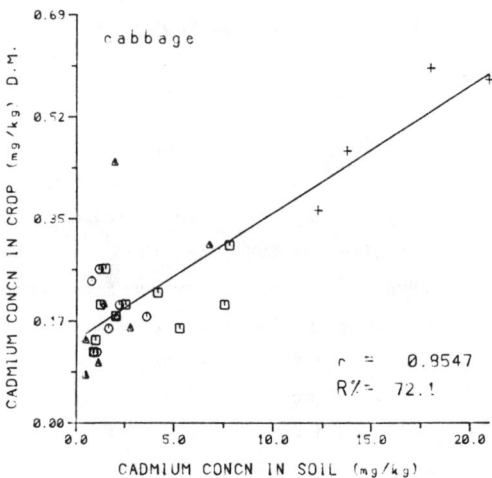

4. DISCUSSION

In Table II, cadmium concentrations observed in the crops as reported in Fig. 2 have been converted to give values for potential human dietary intake of cadmium. The results used for this purpose relate only to the high-metal sludges (S_1 and S_4). Not all the crops eaten by man were grown in the experiment and this has required grouping of crops according to a league table of crop sensitivity to cadmium devised by Davis and Carlton-Smith[3]. It is thought that the grouping used overestimates the potential uptake which would have occurred had results for all the individual component crops been available. Food consumption data for average consumers was obtained from statistics published by the Ministry of Agriculture, Fisheries and Food[4]. The results shows that small increases in the cadmium content of staple foods such as potatoes and particularly wheat have a substantially more profound effect on potential dietary intake of cadmium than larger increases in concentration of the element in crops like lettuce. This would apply even to those who eat several times the normal amount of lettuce. In view of the importance of wheat it is of interest to take account of milling. Results (to be published) have shown that flour contains only 57 per cent of the cadmium concentration seen in the wheat grain from which it is made. Thus potential dietary intake is reduced if an individual takes his/her wheat as flour instead of whole grains. No account is taken in the model of the influence of enhanced soil concentrations of cadmium on meat and dairy products for human consumption. It was felt that these are insensitive to increases in the cadmium content of soil except perhaps for offal (liver, kidney) which composes only about 0.5 per cent of the diet on a fresh weight basis[4]. Fig. 3 plots soil concentrations of cadmium against potential dietary intake of cadmium for the average consumer taking all his/her crops from sludge-treated soil. There is a linear relationship between cadmium in soil and potential dietary intake of cadmium. The gradient of the slope is greater if cereals are eaten as whole grain rather than flour (Fig. 3).

One basis for an objective soil concentration limit for cadmium would be that concentration which produces a potential dietary intake of 70 µg/day, the value calculated independently by both the World Health Organisation (WHO) and the United States Environmental Protection Agency (EPA) as being the maximum acceptable daily intake of cadmium. For the whole grain diet this value is generated by a soil concentration of 6.24 mg Cd/kg and for the flour diet the soil concentration is 12.90 mg Cd/kg. Tolerable accuracy of the model is suggested by the finding that unsludged soil could generate a dietary intake of about 20 µg/day, close to the

Table II. Potential dietary intake of cadmium by an average consumer taking all his/her crops from soil treated with sludge in the Royston experiment (data for sludges S_1 and S_4 only)

Crop	Daily intake (g) Fresh*	Dry	Daily intake of cadmium (μg) from crops according to treatment S_1R_0	S_1R_1	S_1R_2	S_1R_3	S_1R_4	S_4R_4	S_4R_5
Potatoes	187.5	37.6	4.51	8.27	6.76	8.64	8.27	6.76	6.76
Cabbage other green vegetables	56.0	4.39	0.61	0.88	0.83	0.83	1.32	1.80	2.59
Lettuce and other green salad	5.3	0.17	0.11	0.20	0.23	0.23	0.30	0.25	0.33
Red beet, other root vegetables, onions, leeks and tomatoes	70.7	9.19	1.56	2.75	2.94	4.32	5.14	4.96	5.52
Cereals-whole grain or flour	226.4	194.7							
Whole grain			17.52	36.99	42.83	48.68	64.26	52.57	101.25
Total	545.9	246.1							
Whole grain			24.31	49.09	53.59	62.70	79.29	66.34	116.45
Flour			16.78	33.19	35.18	41.77	51.67	43.74	72.93

*data from reference 4

Fig. 3. The relationship between cadmium concentrations in soil and
potential dietary intake of cadmium for an average consumer
taking all his/her crops from sludge-treated soil

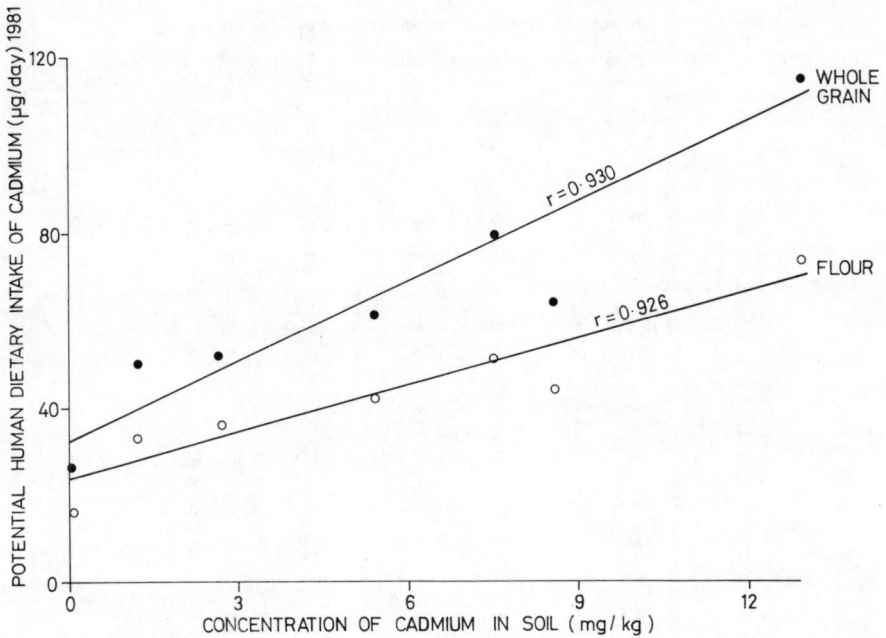

estimated value for UK citizens[5]. On this basis, it is suggested that a
soil concentration in the range 6.0-12.0 mg Cd/kg (equivalent to a loading
rate of 10-22 kg Cd/ha) is acceptable for calcareous soils (pH value 7-8)
receiving sludge. It is our intention to improve and consolidate the
model used to arrive at this value but it has the following built-in
safety factors:

a. Only 1.27 per cent of agricultural land in the UK receives sludge
each year. The average dressing increases the soil concentration of
cadmium by about 0.025 mg/kg/yr. It is almost inconceivable that any
individual could live on crops taken wholly from sludge-treated soil.

b. Whilst the model is based on an average consumer (who, in reality,
does not exist), an unusual diet consisting for instance of five
times the normal intake of lettuce would have little effect on
dietary intake of cadmium. Wheat is much more important but few
people grow their own supplies. This crop is usually collected and
stored centrally so any contaminated sample would be diluted out.
The diet of most individuals would include at least some refined grain
(flour) of lower cadmium content.

c. The WHO/EPA maximum acceptable daily intake of cadmium of 70 μg/day is not a toxic threshold but has its own safety factor. It is estimated that 200 μg/day of cadmium sustained over a 50 year period would be needed to produce kidney damage (of doubtful clinical significance) in the most sensitive individual. (See references 6 and 7.)

5. CONCLUSIONS

For an average consumer taking all his/her crops from sludge-treated soil, a cadmium concentration in soil (pH value 7.1-8.2) of 6.0-12.0 mg/kg would be compatible with the WHO/EPA maximum acceptable dietary intake of cadmium of 70 μg/day. Interpretation of this kind, based on data collected from field trials, can serve to provide an objective basis for soil concentration limits where sewage sludge is used on agricultural land. Current UK guidelines recommend a maximum soil cadmium concentration of 3.5 mg/kg for arable land (pH value 6.5) receiving sludge, to be approached gradually over a 30 year period[1].

ACKNOWLEDGEMENTS

The field trial at Royston is being financed by the UK Department of the Environment as part of its comprehensive research programme on the disposal of sewage sludge. (DOE Contract 480/512)

REFERENCES

1. DEPARTMENT OF THE ENVIRONMENT/NATIONAL WATER COUNCIL. Report of the Sub-Committee on the Disposal of Sewage Sludge to Land. DOE/NWC Standing Technical Committee Report 20. NWC, London, 1981.

2. DAVIS, R.D. and STARK, J.H. Paper presented to the Second European Symposium on Characterization, Treatment and Use of Sludge, Commission of European Communities Conference, Vienna, October 1980. pp. 687-698, D. Reidel, Dordrecht.

3. DAVIS, R.D. and CARLTON-SMITH, C.H. Crops as indicators of the significance of contamination of soil by heavy metals. Water Research Centre Technical Report TR140, 1980.

4. MINISTRY OF AGRICULTURE, FISHERIES AND FOOD. Household Food Consumption and Expenditure 1979. Annual Report of the National Food Survey Committee HMSO, London, 1981.

5. DEPARMENT OF THE ENVIRONMENT. Cadmium in the Environment and its Significance to Man. Pollution Paper 17. HMSO, London, 1980.

6. DAVIS, R.D. and COKER, E.G. Cadmium in Agriculture with Special Reference to the Utilisation of Sewage Sludge on Land. Water Research Centre Technical Report TR139, 1980.

7. NAYLOR, L.M. and LOEHR, R.C. Increase in dietary cadmium as a result of application of sewage sludge to agricultural land. Environmental Science and Technology 1981, 15, 881-886.

SOIL-CHEMICAL EVALUATION OF DIFFERENT EXTRACTANTS FOR HEAVY METALS IN SOILS

D. R. SAUERBECK and E. RIETZ

Institute of Plant Nutrition and Soil Science
Federal Research Center of Agriculture Braunschweig-Völkenrode (FAL)
Federal Republic of Germany

Summary

25 different extractants were tested concerning their capability to dissolve heavy metals from three contaminated soils. The results indicate different binding forms and a strongly pH-dependent solubility of Cd, Zn and Pb. Cd has always been more mobile than Zn or Pb, and the dissolution of heavy metals from mineral soils was considerably more than from sludge. Alkaline solutions hydrolized only little organic-matter bound Zn and Pb but no Cd. Mg and CaCl2 proved to be more efficient extractants than nitrates or neutral acetate salts. Only Zn was attacked by ammonium oxalate. Among the different salt/acid mixtures the lactate-containing reagents gave the highest recoveries. Acid complexing EDTA-mixtures dissolved more than neutral DTPA-containing solutions. Strong acids tended to blur all the solubility differences. Fractionating extractions and wide-range acidifications were laborious and somewhat contradictory. Best results could be obtained by weakly acid but well buffered extracting solutions.

1. INTRODUCTION

The chemical forms and the binding of heavy metals in soils depend on the physico-chemical conditions prevailing. Part of them exist as insoluble minerals such as oxihydrates, carbonates or phosphates, others are fixed or adsorbed by the soil colloids, and a certain portion also appears to be complexed by the organic matter (5, 7, 8, 9, 12, 32). It is clear that these different forms of binding decide upon both the mobility of such compounds in soils and upon their plant availability. Accordingly, more should be known about the forms and behaviour of heavy metals in soils in order to enable their environmental significance to be correctly evaluated (6, 7, 34).

Since a direct separation and identification of heavy metal compounds is almost impossible for such complex systems as soils, a number of investigators have tried to separate them indirectly into groups of different solubility by means of specific extractants (e. g. 1, 2, 4, 5, 10, 11, 13, 14, 20, 21, 24, 27, 30, 33). They also took into consideration their pH- and rh-dependent chemical behaviour (3, 5, 9, 15, 16, 17, 18, 19, 31). However, the knowledge in this field is still far from being complete, because most of the published data can neither be generalized nor directly compared. It is for this reason that in the present investigation a large number of different extracting and fractionation procedures has been applied to compare the behaviour of Cd, Zn and Pb in three contaminated soils.

2. MATERIALS AND METHODS

Two of the investigated soils came from an area near to a large zinc smelter, and the third one was formed during a 6 - 8 year's decomposition period of the solids dredged from a sewage sludge pond. As Table I shows, soils No. 1 + 2 are of similar composition as fas as grain size, CEC and

Table I: Chemical characterization and heavy metal contents of the soils used

No. and origin	soil fract.	CEC mval/100g	pH 0.02m CaCl$_2$	C$_{org}$ %	CaCO$_3$ %	<2 μm /%	Cd[x) mg/kg	Zn[x) mg/kg	Pb[x) mg/kg
1	<2 mm	14	7.2	1.8	2.6	17	34	1994	951
field soil 0 - 25 cm	<2 μm	44		2.0	3.4	-	70	6210	2540
2	<2 mm	12	5.6	1.9	0.3	13	105	5522	2120
field soil 0 - 25 cm	<2 μm	42		3.5	0.8	-	244	12560	4410
3 decomp. sludge	<2 mm	31.6	6.3	7.9	4.6	5	44	1512	138

[x)]extraction with aqua regia

organic matter are concerned, but differ in pH and carbonate percentage and particularly in their heavy metal content. The sludge-derived soil No. 3 contains little clay but a considerable amount of organic matter and heavy metals. The major portion of these heavy metals in the soils No. 1 and 2 were found to be concentrated in their clay fraction.

A total of 25 chemical extractants as recommended for the extraction of trace elements and heavy metals from soils (e. g. 2, 4, 7, 34) was selected according to Table II. In each case 100 ml solution was added to 10 g of air dry soil, the mixture was sonicated for 3 minutes and afterwards shaken for a period of 2 hours. Cd, Zn and Pb in the filtrates were determined in the acethylene flame of a Perkin-Elmer atomic absorption spectrometer 560.

Table II: Composition, concentration and pH-value of the extracting solutions

A. Dilute alkalines	pH
(1) 0.1 N NaOH	13.1
(2) 1.0 N NaOH	13.7
B. Salt solutions	
(3) 0.1 N NaNO$_3$	6.0
(4) 1.0 N NaNO$_3$	5.7
(5) 0.1 N Ca(NO$_3$)$_2$	5.8
(6) 1.0 N Ca(NO$_3$)$_2$	5.2
(7) 0.1 N CaCl$_2$	5.7
(8) 1.0 N CaCl$_2$	5.8
(9) 2.0 N CaCl$_2$	5.8
(10) 2.0 N MgCl$_2$	6.8
(11) 0.55 N (COO)$_2$(NH$_4$)$_2$	3.3
(12) 1.0 N (CH$_3$COO)NH$_4$	7.0
C. Salt/acid-mixtures	
(13) 0.03 N NH$_4$F/0.1 N HCl (1 : 1)	1.8
(14) 0.1 N (CH$_3$CHOHCOO)NH$_4$/0.4 N CH$_3$COOH	3.7
(15) 0.1 M (CH$_3$CHOHCOO)$_2$Ca/0.1 M (CH$_3$COO)$_2$Ca/0.3 M CH$_3$COOH	4.2
(16) 0.5 N (CH$_3$COO)NH$_4$/0.5 N CH$_3$COOH	4.7
D. Complexing agents	
(17) 0.005 M DTPA/0.1 M TEA/0.01 M CaCl$_2$	7.3
(18) 0.005 M DTPA/1.0 M NH$_4$HCO$_3$	7.8
(19) 0.05 N (NH$_4$)$_2$XDTA	4.3
(20) 0.5 N (CH$_3$COO)NH$_4$/0.02 M XDTA	4.1
E. Dilute acids	
(21) 2.0 N CH$_3$COOH	2.0
(22) 0.1 N HCl	0.9
(23) 0.1 N HNO$_3$	0.9
(24) 0.5 N HNO$_3$	0.2
(25) 2.0 N HCl	0.0

Additional step-wise extractions were made according to Figure 1 by subsequent treatments of 10 g soil with 50 ml solutions in the sequence deionized water, 1 N ammonium acetate and 0.05 N Cu-acetate (26, 27, 29). Each of these extractions was repeated twice, and the centrifugates combined for the analyses.

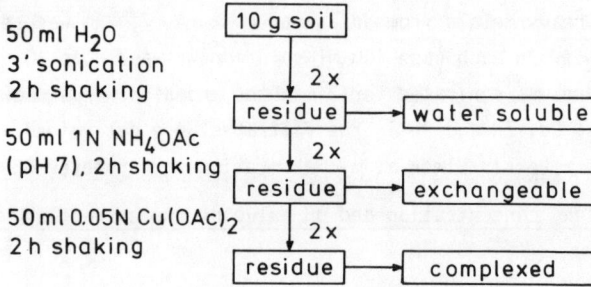

Figure 1: Step-wise extraction of heavy metals from soil according to SMITH and SHOUKRY (1968)

A fractionating electrodialysis was carried out by exposing 10 g soil to a constant potential of 27.5 V/cm according to Figure 2 (22, 35, 36). The soil was separated from the electrode chambers by cellophane membranes

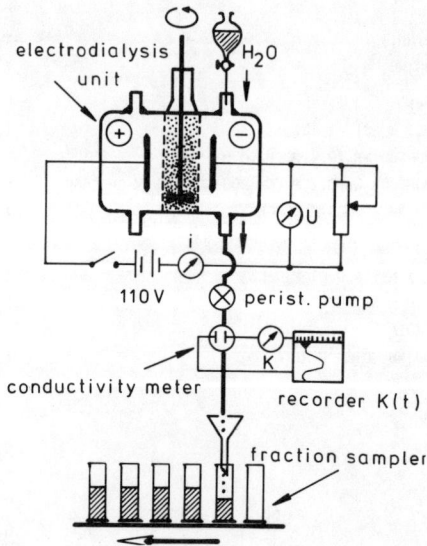

Figure 2: Fractionating electrodialysis of soil samples (schem.)

and continuously suspended by stirring. Samples were automatically drawn from the cathode liquid at a rate of about 3.6 ml/min and collected in 10 ml fractions after passing a conductivity meter. A total of 50 fractions was taken, corresponding to an overall dialysis-time of 2 hours and 15 minutes.

The pH-dependent dissolution of the three heavy metals was studied according to COTTENIE (4) by suspending 20 g soil in 80 ml H_2O. These suspensions were then acidified with 0.5 N HNO_3 and held at the desired acidity levels ranging between pH 1.0 and the original soil-pH for 1 hour. Afterwards the volumes were made up to 100 ml immediately and the solutions filtered.

Another set of pH-denpendent extractions was made using Na-acetate solutions buffered with acetic acid to pH-values ranging between 3.5 and 5.5. As in the other treatments (except the electrodialytic fractionation and the potentiometric acidification), 10 g soil was shaken for 2 hours with 100 ml solution after a 3 minute sonication pretreatment.

3. RESULTS AND DISCUSSION

3.1 COMPARISON OF DIFFERENT EXTRACTANTS

The results of all the extractions are summarized in Figure 3. The ordinate represents the relative solubilities as percent of the total heavy metal contents extracted by aqua regia (see Table I), while on the abscissa the identification number of each particular extractant is shown (see Table II).

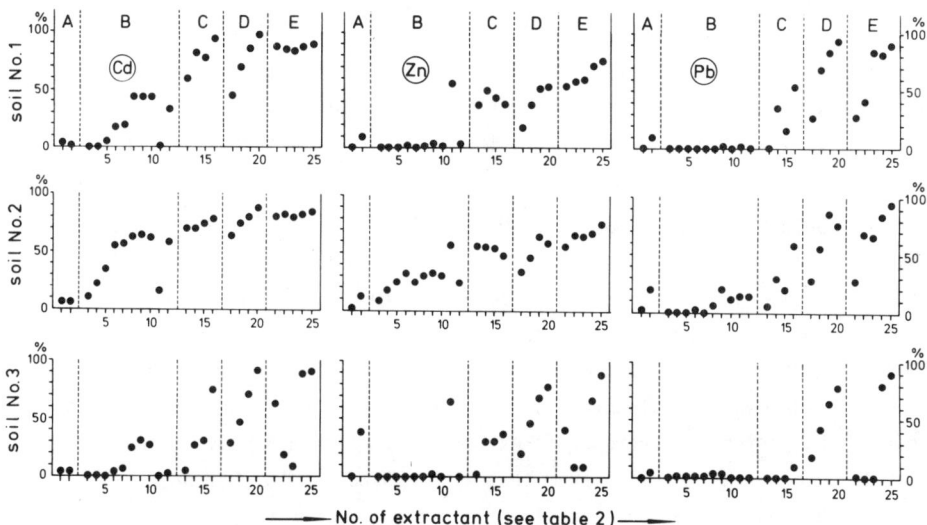

Figure 3: Solubility of heavy metals from contaminated soils in 25 different extractants (relative values in % of aqua regia)

The vertical dotted lines separate the different groups of extractants according to their particular chemism into A) dilute alkalies, B) salt solutions, C) salt/acid-mixtures, D) complexing agents, and E) dilute acids. The three blocks of data from left to right presented the three elements Cd, Zn and Pb, while the different soils have been arranged in the vertical sequence.

Although at the first glance this graph looks somewhat confusing, it shows very pronounced differences in the dissolution of the three elements depending on both the types of extractants used and on the soils being studied. As an example, soil No. 2 in the center of Figure 3 not only contains the largest total amounts of heavy metals (see Table I) but also exhibits the highest relative solubilities in most of the extractants compared. On the other hand, most of these extractants except some complexing agents and the strong acids appear to be rather inefficient when applied to the sludge-derived soil No. 3.

As regards to the element <u>cadmium</u>, both alkali concentrations (group A, No 1 + 2) did not extract very much. This could be attributed simply to its precipitation at the high pH-value. However, one may also speculate that not much Cd had been incorporated into the alkali-soluble soil organic matter fraction.

From the salts (group B) 0.1 and 1 N $NaNO_3$ (No. 3 + 4) has been suggested as a promising extractant (12) but dissolved only traces except from the highly contaminated soil No. 2, where between 10 and 20 % of the Cd were brought in solution. - Since Ca-ions are usually preferred for soil testing purposes, the extracts (No. 5 + 6) represent 0.1 and 1 N $Ca(NO_3)_2$ solutions, which in fact extracted more than the corresponding $NaNO_3$-concentrations did, and again corroborated the higher relative solubility of the Cd in the more acid soil No. 2.

As could be expected, 0.1 N, 1 N and 2 N $CaCl_2$ (No. 7 - 9) proved to be more effective than the corresponding Na- and Ca-nitrates were. According to general chemical knowledge this may be due to the formation of cadmium chloro-complexes, although a concentration increase from 1 N to 2 N $CaCl_2$ (No. 8 + 9) did not make big differences any more. Nevertheless, the differences between NO_3- and Cl-extracts were rather pronounced and should be kept in mind when selecting appropriate extractants. - The 2 N $MgCl_2$ extract (No. 10) did not differ much from that with 2 N $CaCl_2$ (No. 9). NH_4-oxalate (No. 11) turned out to be rather inefficient for Cd contrary to the Zn which will be described later. This was not the case with the neutral NH_4-acetate (No. 12), which at least in the soils No. 1 + 2 turned out to be about as efficient as the $CaCl_2$-solutions were. It is not quite clear why this did

not apply to the sludge-derived soil No. 3, but the behaviour of this particular soil against all the salts tested was consistently different from the
other two soils.

This has also been found for the salt/acid-mixtures (group C), of
which the NH_4F/HCl-solution (No. 13) extracted about 60 % of the Cd from the
soils No. 1 + 2 but much less from the soil No. 3. The more strongly buffered
ammonium lactate/acetic-acid mixture (No. 14) as well as the corresponding
Ca-compounds (No. 15) removed even more of the Cd from all the three soils,
and the conventional ammonium acetate/acetic-acid mixture (No. 16) even
exceeded the efficiency of the former extractant. This means that most of the
low-pH buffered acid agents suggested for routine soil analysis can be expected to dissolve very much of the cadmium present.

As far as the complexing agents (group D) are concerned, their
pH-value also appears to be of the utmost importance. The slightly alkaline
DTPA mixtures (No. 17 + 18), therefore, turned out to be inferior as compared
with the NH_4-EDTA (No. 19) and of course the more concentrated ammonium-
acetate/EDTA mixture (No. 20). However, even the neutral DTPA agents (No. 17
+ 18) dissolved rather much of the Cd from the otherwise fairly resistant
soil No. 3, and the EDTA admixtures (No. 19 + 20) resulted in very high
dissolution percentages from all the three soils tested.

Due to their strong acidity all the dilute acids (group E, No. 21 -
25) also extracted comparatively large percentages of the Cd present, except
for the sludge-derived soil No. 3, where the high contents of carbonate and
organic matter reduced the efficiency of the 0.1 N HCl and HNO_3 (No. 22 + 23).
This means that unbuffered acids should not be used as heavy metals extractants from soils because of their unrealistically large recoveries and their
unpredictable extent of neutralization during the extraction procedure.

The reaction of all these extractants with zinc can be discussed more
briefly, because of its similarity with the cadmium already described. Of
the dilute alkalies (group A) only the 1 N NaOH (No. 2) resulted in
some dissolution of Zn. As this extractability was most pronounced in soil
No. 3, it may be concluded that part of this Zn there is bound by its organic
matter. Contrary to the findings with Cd, of the salts tested (group B)
NH_4-oxalate (No. 11) turned out to be rather effective because it dissolved
about 60 % of the total Zn. This is amazing because only 20 - 30 % of the Fe
was simultaneously extracted. The Ca- and Mg-chlorides (No. 7 - 10) were much
less effective for Zn than for Cd, which again suggests a particular reaction
of the chloride ions with the cadmium present.

The extraction efficiencies of the salt/acid mixtures (group C), the
complexing agents (group D) and the dilute acids (group E) were

essentially similar to what has been found for the Cd. However, the proportion of Cd dissolved was in most cases higher than that for Zn. This suggests
a more strong binding of Zn in the soil as compared with the Cd, which is
important especially since the amounts of Zn were so very much higher. As a
probable consequence of their differing pH it may by also noted that the two
DTPA-extractants (No. 17 + 18) suggested for Zn by LINDSAY and NORVELL (21)
and SOLTANPOUR and SCHWAB (30) extracted much less of the Zn than the more
acid EDTA-containing reagents (No. 19 + 20) used by LAKANEN and ERVIOE (20)
did. This does not necessarily imply that the quantitatively more efficient
extractant is always the more representative one as far as the availability
to plants is concerned, but it should be at any rate known.

    Although at a lower proportion than that of the Zn, a part of the ele-
ment lead could also be dissolved by 1 N   NaOH (group A, No. 2). Taking
into consideration the large stability of mineral lead compounds at high pH-
levels, the most reasonable explanation for this NaOH-solubility is the partial binding of Pb by soil organic matter.

    With some exceptions in soil No. 2, most of the   salt solutions
(group B) and one of the   salt/acid mixtures   (group C, No. 13) did not
extract any appreciable amounts of Pb. Limited exceptions occured with the
neutral NH_4OAc (No. 12) and with 2 N salt concentrations (No. 9 + 10), which
was presumably due to a more effective ion exchange. A pronounced dissolution
of Pb, however, occured by the low-pH buffered ammonium-lactate/acetic-acid
and the NH_4OAc/acetic-acid mixtures (No. 14 + 16), although this was not the
case in the carbonate-rich soil No. 3 and with the corresponding calcium
salt compounds. (No. 15).

    The complexing agents (group D) behaved against Pb rather similarly
as against Zn, showing a clear-cut pH-dependence (No. 17 - 20) but a fairly
high lead solubility even at neutral pH-levels due to the presence of DTPA
(No. 17 + 18). The EDTA-extractability of the lead (No. 19 + 20) even exceeded that of the Zn, which shows that a lot of this elements had been complexly bound in these soils.

    The solubility of soil Pb in the dilute acids (group E) was not so
much different from that of Zn, except that the low concentrated mineral
acids (No. 22 + 23) and the less dissociated acetic acid (No. 21) behaved relatively poor. It just seems to require a sustained pH-value below 2.5 in order
to get an appreciable amount of soil Pb in solution if no additional complexing agents are present.

3.2 STEP-WISE EXTRACTION OF FRACTIONS

    In contrast to all the single extractants mentioned before, the different

forms of binding for heavy metals in soils could be shown much more clearly
by their step-wise extraction (25, 26, 29). As can be seen from Figure 4,
less than 4 % of the Cd and Zn had been dissolved in water, whereas the soil
lead was nearly water insoluble at all. The extraction results with neutral
NH_4-acetate have been already shown before. These exchangeable portions are
much higher for Cd than for Zn, and are fairly low for Pb. A similar sequence
applies to the complexed portions, except for soil No. 1 where this Zn fraction
even predominated. The same phenomenon has been found for the lead in soil No. 3.

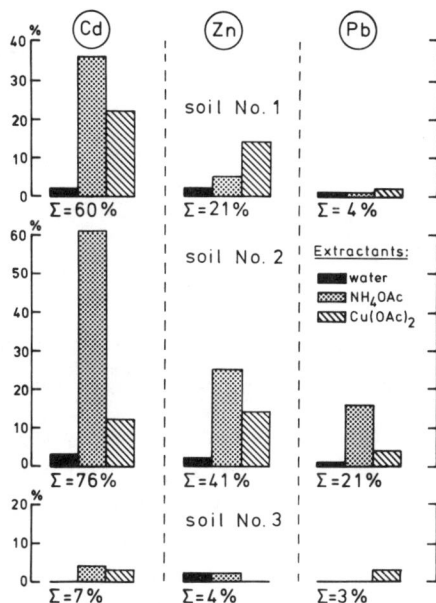

Figure 4: Results of step-wise extraction for heavy metals from soils
          (relative values in % of aqua regia)

As to these different soils, it is obvious that the relative solubility
of their heavy metals decreased with increasing carbonate and organic matter
content. By far the lowest solubility has been found for the sludge-derived
soil No. 3. The proportion of its total Cd, Zn and Pb released by all three
extractions was 7, 4 and 3 % respectively, as compared with 60, 21 and 4 %
or with 76, 41 and 21 % in the two mineral soils No. 1 + 2. Consequently, it
can not be expected that one particular extractant would be equally effective
in a whole range of different soils.

3.3 FRACTIONATING ELECTRODIALYSIS

Another interesting approach to distinguish between heavy metal fractions
of different stability might be the electro-ultrafiltration as it is occasion-
ally used for soil testing (23, 28). Before applying this method to the

analysis for heavy metals in soils, it has first been proven that dissolved Cd-, Zn- and Pb-salts show a most similar behaviour in an electrodialytic cell, i. e. their concentration gradients versus time almost coincide (25).

However, Figure 5 shows that if the soil No. 2 was electrodialyzed in the same way, the three heavy metal peaks appeared considerably later than the bulk of the other ions, and showed a characteristic sequence (25). The first

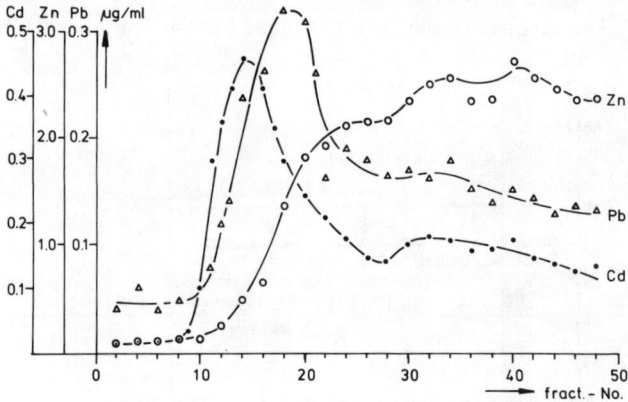

Figure 5: Fractionating electrodialysis of heavy metals from soil No. 2

distinct peak to be seen was that of the Cd, followed by the Pb and consider-ably later by a much flatter and more extended peak of the Zn. Opposite to this differing velocity of liberation and movement in the electric field, the average concentrations in the cathode liquid were equally low for Cd and Pb and about 9 times higher for Zn, in contrast to the total contents of Cd, Zn and Pb which were of the ratio 1 : 55 : 21. This means that electrodialysis can characterize a time sequence in the appearance of soil heavy metals. However, only the average concentration of heavy metals electrodialytically removed is actually correlated with their chemical extractability.

3.4 pH-DEPENDENT DISSOLUTION

This is one of the reasons why COTTENIE (4) suggested to use a defined acidification with HNO_3 and subsequent extraction at controlled pH-values in order to characterize the stability of heavy metals in soils. As can be seen from Figure 6 both Cd and Zn became mobile at much lower H^+-ion concentra-tions than the Pb did, which corresponds to the former extraction results already discussed. However, here again the sludge-derived soil No. 3 differed from the two mineral soils because the proportion dissolved was less at comparable pH-values, and the Cd was less soluble than the Zn in this partic-ular soil. According to current knowledge this can be attributed to the large amounts of organic matter present.

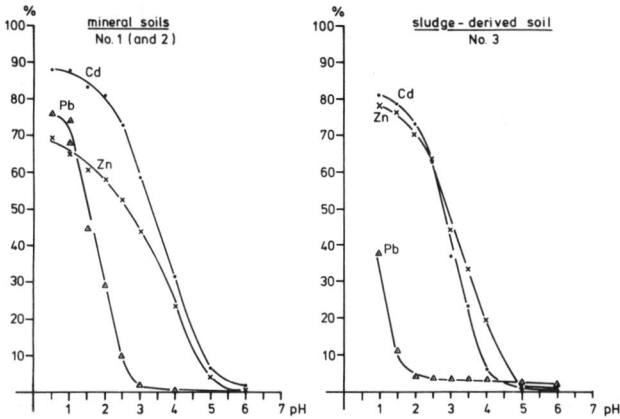

Figure 6: Heavy metal solubility after defined acidification and soil
extraction at controlled pH-values according to COTTENIE (1977)
(relative values in % of aqua regia)

The most different ratios between the dissolved proportions of the three elements
for the different pH-values make clear that extractants with pH-values lower
than pH 3 - 4 can not be recommended, because there even the most inert element Pb gets in solution, and the extraction efficiency for Cd and Zn becomes
unrealistically high. Such pH-dependent dissolution curves according to
COTTENIE may therefore be useful to characterize fundamental differences
between soils, but will certainly not become a conventional test method even
if it were not as laborious as it actually is.

The idea of maintaining the pH during the heavy metal extractions at a
fixed value is nevertheless appealing, but could be done much more simply by
adjusting an acetate/acetic-acid buffer solution accordingly. This has been
done using an NaOAc/acetic-acid system for obtaining the informations summarized in the last Figure 7. The most striking result from this treatment is
that here all three soils differed greatly in their pH-dependent heavy metal
mobility, quite in contrast to the pH-adjusted HNO_3-extractions according to
COTTENIE (4) mentioned before (Figure 6). There only the sludge-derived soil
No. 3 differed whereas the mineral soils No. 1 + 2 behaved almost equal.
Here we have found distinct differences among all three soils, and in all
cases a considerably lower extraction efficiency of the NaOAc-buffer as
compared with the NHO_3.

- 157 -

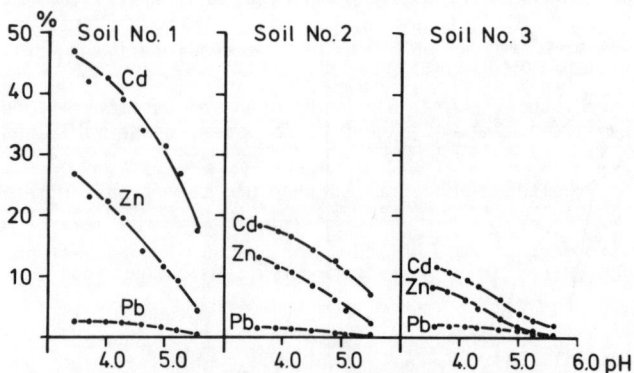

Figure 7: Heavy metal solubility in NaOAc buffered with acetic acid to different pH-values (% of aqua regia)

As far as the ratios between the NaOAc-extracted proportions of Cd, Zn and Pb are concerned, their changes between pH 5.5 and 3.5 according to Figure 7 appear to be small. Hence, coming back to the initial comparison of 25 different extractions, the conclusion suggests itself that the weakly acid and well buffered salt/acid mixtures will be the most desirable extractants for heavy metals from soils. Referring to what is already known from conventional soil testing, this could in fact have been expected, but the number of extractants and methods suggested in the literature still vary considerably. The question which particular pH-value should be preferred can only be tackled if more experimental comparisons with the true plant uptake are available.

4. LITERATURE

(1) ANDERSSON, A.: Relative efficiency of nine different soil extractants. Swedish J. Agric. Res. 5, 125 - 135, 1975.

(2) BAUER, A.: Considerations in the development of soil tests for "available" zinc. Soil Sci. and Plant Anal. 2, 161 - 193, 1971.

(3) BRÜMMER, G.: Einfluß organischer Substanzen auf Löslichkeit, Bindung und Umwandlung mineralischer Bodenkomponenten. Mitt. Dt. Bodenkdl. Ges. 27, 173 - 180, 1978.

(4) COTTENIE, A.: Content and behaviour of trace elements in soils. FAO-Meeting Bonn, Network on trace element studies, 15 - 18 March, 15 pp., 1977.

(5) COTTENIE, A.: Soil and plant testing as a basis of fertilizer recommendation. FAO, Rome, Special Report, 100 pp., 1978.

(6) COTTENIE, A., CAMERLYNCK, R., VERLOO, M., VELGHE, G., KIEKENS, L. and DHAESE, A.: Essential and non-essential trace elements in the system soil - water - plant. Labor. Analytical - Agrochem. State University, Ghent-Belgium, 75 pp., 1979.

(7) COX, F. R. and KAMPRATH, E. J.: Micronutrient soil tests. In: Mordtvedt, J. J., Giordano, P. M., Lindsay, W. L. (Eds.): Micronutrients in agriculture. Proc. Symp. Muscle Shoals, Alabama, April 20 - 22, 1971. Soil Sci. Soc. Am., Inc., Madison, 289 - 317, 1972.

(8) FASSBENDER, H. W. und SEEKAMP, G.: Fraktionen und Löslichkeit der Schwermetalle Cd, Co, Cr, Cu, Ni und Pb im Boden. Geoderma 16, 55 - 69, 1976.

(9) FÖRSTNER, U.: Umweltchemische Untersuchung und Bewertung von metallkontaminierten Schlämmen. Chemiker-Zeitg. 105, H. 6, 1 - 10, 1981.

(10) FRANCK, E. v.: Vergleich von Methoden zur Bestimmung des verfügbaren Zinks im Boden. Mitt. Dt. Bodenkdl. Ges. 29, 595 - 602, 1979.

(11) GHANEM, A., KEILEN, K. und STAHR, K.: Freisetzung und Mobilität von Spurenelementen in Braunerden und Podsolen des Bärhaldegranitgebietes. Mitt. Dt. Bodenkdl. Ges. 29, 577 - 586, 1979.

(12) GUPTA, S. and HÄNI, H.: Easily extractable Cd-content of a soil - its extraction, its relationship with the growth and root characteristics of test plants and its effect on some of the soil microbiological parameters. Proc. of II. Europ. CEC Symp. on characterization, treatment and use of sewage sludge, Vienna, October 21 - 23 1980, 665 - 676, 1981.

(13) HAQ, A. U., BATES, T. E. and SOON, Y. K.: Comparison of extractants for plant-available zinc, cadmium, nickel, and copper in contaminated soils. Soil Sci. Soc. Amer. J. 44, 772 - 777, 1980.

(14) HAVLIN, J. L. and SOLTANPUR, P. N.: Evaluation of the NH_4HCO_3/DTPA soil test for iron and zinc. Soil Sci. Soc. Amer. J. 45, 70 - 75, 1981.

(15) HERMS, U. und BRÜMMER, G.: Der Einfluß des pH-Wertes auf die Löslichkeit von Schwermetallen in Böden und Komposten. Mitt. Dt. Bodenkdl. Ges. 25, 139 - 142, 1977.

(16) HERMS, U. und BRÜMMER, G.: Löslichkeit von Schwermetallen in Siedlungsabfällen und Böden in Abhängigkeit von pH-Wert, Redoxbedingungen und Stoffbestand. Mitt. Dt. Bodenkdl. Ges. 27, 23 - 34, 1978.

(17) HERMS, U. und BRÜMMER, G.: Einfluß organischer Substanzen auf die Löslichkeit von Schwermetallen. Mitt. Dt. Bodenkdl. Ges. 27, 181 - 192, 1978.

(18) HERMS, U. und BRÜMMER, G.: Einfluß der Redoxbedingungen auf die Löslichkeit von Schwermetallen in Böden und Sedimenten. Mitt. Dt. Bodenkdl. Ges. 29, 533 - 544, 1979.

(19) HERMS, U. und BRÜMMER, G.: Einfluß der Bodenreaktion auf die Löslichkeit und tolerierbare Gesamtgehalte an Nickel, Kupfer, Cadmium und Blei in Böden und kompostierten Siedlungsabfällen. Landw. Forsch. 33, 408 - 423, 1980.

(20) LAKANEN, E. and ERVIOE, R.: A comparison of eight extractants for the determination of plant available micronutrients in soils. Acta Agr. Fenn. 123, 223 - 232, 1971.

(21) LINDSAY, W. L. and NORVELL, W. A.: Development of a DTPA soil test for zinc, iron, manganese, and copper. Soil Sci. Soc. Amer. J. 42, 421 - 428, 1978.

(22) MANEGOLD, E.: Über Kapillarsysteme, XVIII/1. Technische und apparative Neuerungen bei der Darstellung, Kennzeichnung und Anwendung von Membranen. Koll.-Z. 78, 129 - 148, 1937.

(23) NEMETH, K.: The availability of nutrients in the soil as determined by electro-ultrafiltration (EUF). Advanc. Agron. 31, 155 - 188, 1979.

(24) NEUHAUSER, E. F. and HARTENSTEIN, R.: Efficiencies of extractants used in analysis of heavy metals in sludges. J. Environ. Qual. $\underline{9}$, No. 1, 21 - 22, 1980.

(25) RIETZ, E. und SÜCHTIG, H.: Extraktionsverhalten und Bindung von Schwermetallen in Böden unterschiedlichen Belastungsgrades. Landw. Forsch. 38. Sh. (in press), 1982.

(26) SEDBERRY, J. E., Jr., and REDDY, C. N.: The distribution of zinc in selected soils in Louisiana. Commun. Soil Sci. Plant Anal. $\underline{7}$, 787 - 795, 1976.

(27) SEDBERRY, J. E., Jr., MILLER, B. J. and SAID, M. B.: An evaluation on chemical methods for extracting zinc from soils. Commun. Soil Sci. and Plant Anal. $\underline{10}$, 689 - 701, 1979.

(28) SINCLAIR, A. H.: Desorption of cations from Scotish soils by electro-ultrafiltration. J. Sci. Food Agric. $\underline{31}$, 532 - 540, 1980.

(29) SMITH, R. L. and SHOUKRY, F. S. M.: Zinc uptake and growth of field beans as affected by organic additions. In: Isotopes and radiation in soil-organic matter studies. Proc. Symp. IAEA/FAO, Vienna, 3 pp., 1968.

(30) SOLTANPOUR, P. N. and SCHWAB, A. P.: A new test for simultaneous extraction of macro- and micro-nutrients in alkaline soils. Commun. Soil Sci. and Plant Anal. $\underline{8}$, 195 - 207, 1977.

(31) SOMMERS, L. E. and LINDSAY, W. L.: Effect of pH and redox on predicted heavy metal-chelate equilibria in soils. Soil Sci. Soc. Amer. J. $\underline{43}$, 39 - 47, 1979.

(32) VIETS, F. G., Jr.: Chemistry and availability of micronutrients in soils. Agric. and Food Chem. $\underline{10}$, 174 - 179, 1962.

(33) VIRO, P. J.: Use of ethylendiaminetetraacetic acid in soil analysis: I. Experimental. Soil Sci. $\underline{79}$, 459 - 465, 1955.

(34) WALSH, L. M. and BEATON, J. D. (Eds.): Soil testing and plant analysis. Soil Sci. Soc. Amer., Inc., Madison, Wisc. (USA), 488 pp., 1973.

(35) WOLLENBERG, H.: Ultrafiltration, Dialyse und Elektrodialyse. In: Schormüller, J., Handb. Lebensm.-Chem., Bd. 2/Teil 1, 85 - 111, 1965.

(36) YOUNG, P.: An introduction to electrodialysis. J. Soc. Diary Technol. $\underline{27}$, 141 - 151, 1974.

EFFECTS OF CADMIUM HUMATES PREPARED AT pH 4 AND pH 6 ON THE
GROWTH AND MINERAL COMPOSITION OF MAIZE PLANTLETS
CULTIVATED IN NUTRIENT SOLUTIONS

A. GOMEZ

Agronomics Institute, INRA, Bordeaux Research Centre,
33140 Pont de la Maye

Abstract

Maize plantlets at the vegetative stage corresponding to the
unfusting of the third leaf were cultivated in nutrient solutions, some of which contained cadmium in the form of humates
or metallic salts. The experiment was conducted at pH 4 and
pH 6. It was observed that when cadmium is associated with
humic acids, its toxicity is not modified. Cadmium transfer
in the aerial parts of the plant is greatly increased at pH 4
in the presence of humic acids. The phosphorus content of the
plant is inhibited at pH 6 in the presence of cadmium,
especially in mineral form.

I. INTRODUCTION

The enrichment of the soil with heavy metals can be observed throughout the world; it is the results of both human activities and geochemical processes. Such metals that arrive in the soil in this manner can occur in various forms, and their destiny in the soil depends on the characteristics of the latter. It is nevertheless probable that at one time or another a fraction of these metals will occur in ionic form and will therefore be capable of reacting with the argillaceous or organic fraction of the soil. The aim of this work is to examine more particularly the effects of the association of cadmium with humic acids (that can be roughly assimilated with a model of the so-called stable but still reactive fraction of the organic matter in the soil) on the growth and mineral composition of maize plantlets. In order more effectively to control the experimental conditions, the plants were grown in nutrient solutions. The pH values chosen for the experiment correspond to that frequently encountered in the Gascony Landes, a region in which maize is cultivated on a large scale, and to that often obtained after liming (pH 4 and pH 6 respectively).

II. EQUIPMENT AND METHODS

The humic acids are extracted from a soil taken from the surface horizon of a humic podzol developed in sand in the Gascony Landes. This sand contains nearly 4% organic matter and approximately 90% coarse sand.

1. Preparation of the humic acids

Place the sample of soil in a 0.5N solution of caustic soda so that the weight/volume ratio is 1:2. Stir the mixture for two hours and then centrifuge at 4 000 rpm. Filter the supernatant

liquid through a fluted filter paper (Ederol No 13) and adjust
the pH of the filtrate to 1 using a 1/10 solution of pure nitric acid. The suspension obtained is allowed to settle for
24 hours and then centrifuged in order to separate the precipitated humic acids from the fulvic acids remaining in
solution. Suspend the humic acids in double-deionized water and
centrifuge once again; repeat this operation three times in
order to eliminate the nitrate ions and the excess sodium ions.
Suspend the humic acids in double-deionized water and then
place the suspension in a Visking dialyzer with a pore diameter
of 48 Å and dialyse continuously for 24 hours against double-
deionized water. Before use, homogenize the suspension for one
hour with the aid of a magnetic rotating stirrer.

2. Preparation and characterization of the cadmium humates

Divide the suspension of humic acids into two samples and adjust their pH to 4 and 6 respectively using 0.1N caustic soda
and 0.1N nitric acid. The exchange capacity of each of these
samples is determined by the barium chloride method. [Ref. 1]

a) Preparation of the pH 4 cadmium humates

The method developed by Juste et al. [Refs. 2 and 3] is used,
which involves placing the humic acids in contact with a
supersaturated cadmium nitrate solution. The pH is maintained
at a constant value of 4 during the reaction by adding 0.1N
caustic soda. The reaction is complete after 24 hours, and the
compound obtained is purified in the same manner as when the
humic acids were prepared using double deionized water. After
purification the suspension of cadmium humates obtained is
dried at 50°C and the product obtained is finely crushed.

b) Preparation of the pH 6 cadmium humates

In order to prevent the cadmium humates from being contaminated
by cadmium hydroxide, which is formed at pH 6, the dialytic

diffusion method developed by the author [Ref. 1] is used. The
pH is likewise maintained at a constant value of 6 using 0.1N
caustic soda. The reaction is complete after 24 hours, and the
product obtained is purified, dried and crushed as in the case
of the pH 4 substance.

c) Characterization of the cadmium humates and the humic acids

After drying and crushing, the following properties are determined: the dry matter content at 105°C; the ash content at
800°C; the carbon content, with the aid of a Leco carbon
measuring apparatus; the nitrogen content, by the Kjeldahl
process; and the Fe and Cd contents, determined by atomic absorption spectrophotometry (using a Varian 575 spectrophotometer).

The results are given in Table 1 below.

	Dry matter (%)	Ash Content	C (%)	N (%)	C/N	Fe (%)	Cd (%)
pH 4 cadmium humate	93.5	22.1	42.4	2.52	16.8	1.42	14.2
pH 6 cadmium humate	94.4	30.2	34.6	2.28	15.2	1.09	17.3

It can be observed that the exchange capacities of the corresponding humic acids (250 meq/100 g at pH 4 and 425 meq/100 g
at pH 6) have been saturated by cadmium in both cases.

3. Growth test on maize plantlets in different mediums

The plant used for the test is maize of the INRA 260 variety,
which is cultivated in an air-conditioned greenhouse.

a) Preparation of the plantlets and regularization phase

Allow the seeds to germinate in small pots containing fine neutral sand; when the third leaf begins to unfurl, carefully remove the plantlets from the sand and wash their roots repeatedly with double-deionized water in order to remove any remaining sand particles that have adhered to them. Weigh all the plantlets one by one and retain for the rest of the experiment only those that constitute a homogeneous batch. A set of 80 plantlets are thus selected from an initial batch of 200. Each plantlet is placed in an Erlenmeyer flask containing 100 ml of the reference nutrient solution, in which all its roots are immersed. Place the plants in the greenhouse for a further 84 hours, constantly aerating the nutrient solutions by means of a moderate draft of compressed air. This regularization growth phase in intended to eliminate any plants that have suffered from the change of medium. On completion of this phase, once the roots have been rinsed and gently wrung, the plantlets are again weighed in order to form a homogeneous experimental batch of 16 plantlets for each pH value.

At this stage, the pH of the nutrient solution is adjusted to 4 in the case of the plants cultivated in humic compounds prepared at pH 4, and to pH 6 in the case of the other batch. Since these two experiments did not take place simultaneously, batches that were statistically homogeneous but different from one another were obtained for each experiment.

b) Study of the influence of humic compounds on the growth and composition of maize plantlets

The plants are again placed in Erlenmeyer flasks, into which the experimental solutions have been introduced, and kept in a greenhouse for 72 hours, the nutrient medium being aerated in the same manner.

On completion of this final phase, the plantlets are removed, abundantly rinsed with double-deionized water, gently wrung and then weighed. Then separate the aerial parts of the plant from the roots, which are left to dry at 105°C. After drying, the roots and aerial parts are dried and then mineralized (by the dry method at 480°C). Dissolve the residue and determine the cadmium content (by atomic absorption spectrophotometry) and the phosphorus content (by colorimetry of the phospho-molybdic complex that is hot-reduced by ascorbic acid).

The reference nutrient solution consists of a mixture of 200 mg/l $K NO_3$, 100 mg/l $NH_4 NO_3$, 150 mg/l $KH_2 PO_4$ and 100 mg/l $Mg Cl_2$.
The experimental solution contain 10 mg/l of cadmium which is added to the reference nutrient solution in nitrate or humate form. A standard solution is also prepared by adding to the reference solution non cadmium-enriched humic acid in such a concentration that the resulting amounts of nitrogen and carbon are approximately equivalent to those introduced by the cadmium humates.

The pH of the solutions is adjusted to 4 and 6 respectively for the two types of humates; there are therefore four solutions at pH 4 and another four solutions at pH 6.

Each treatment is repeated four times.

III. RESULTS AND DISCUSSION

The results are set out in Table 2 below.

Treatment (repeated on average 4 times)	Dry matter (mg)		P_2O_5 content (%₀ dry matter)		Cd content (ppm dry matter)		
	R	AP	R	AP	R	AP	MI
pH 4							
Mineral sample	93.3	216.7	18.59	18.49	–	–	
Cd $(NO_3)_2$	55.4	159.5	22.13	22.34	1 646	101	6.3
Humic acid	82.2	245.0	21.86	16.84	–	–	
Cadmium humate	63.7	162.6	21.95	21.58	545	59	11.1
pH 6							
Mineral sample	187.9	292.5	15.10	22.40	–	–	
Cd $(NO_3)_2$	119.9	200.1	10.10	18.25	1 823	32	1.76
Humic acid	161.8	282.1	17.50	23.3	–	–	
Cadmium humate	116.9	217.5	13.86	16.43	1 450	29	2

R: roots ; AP: aerial parts; MI: mobility index $= \dfrac{\text{ppm CD AP}}{\text{ppm Cd R}} \times 100$

A. Production of dry matter

On completion of the regularization growth phase, which lasted 84 hours, a statistically homogeneous batch of plantlets was selected. This homogeneity was based both on the fresh weight and the vegetative stage of the plant. This operation was carried out at pH 4 and pH 6.

If the results after 72 hours of growth in the experimental solutions are compared, major differences can be observed.

Statistical processing of the results (the method of contrasts)
highlights the following facts:

- whatever the pH value, the addition of cadmium in salt or
  humate form greatly inhibits the production of dry matter,
  in the case of both the roots and the aerial parts (significant at 1%);

- at both pH 4 and pH 6, no significant effect is observed when
  humic acids are added to the reference nutrient solution;

- the fact that the cadmium is supplied in mineral or humate
  form does not affect this metal's phytotoxicity at either
  pH 4 or pH 6.

B. Effects of the humates on the amount of cadmium and phosphorus absorbed by the plantlets

The statistical interpretation of the results reveals the
following points :

1) Amount of cadmium absorbed

- cadmium accumulates mainly in the roots, irrespective of the
  treatment or the pH;
- at pH 6, the cadmium contents of the aerial parts and the
  roots do not differ significantly according to whether the
  metal is supplied in salt or humate form. The same applies
  at pH 4 to the aerial parts, whereas on the other hand, an
  inhibiting effect on the cadmium content of the roots is
  observed in the presence of humates (significant at 5 %);
- if the mobility indices are considered, with due regard to
  the fact that the experiments at pH 6 and pH 4 were not
  conducted simultaneously, this index appears to drop sharply
  at pH 6 in comparison with pH 4, as a result of the increase
  in the cadmium content of the roots and the decrease in the
  latter in the aerial parts. This effect is observed with
  both the metallic salt and the humates;

- in addition, it is evident that the presence of humic acids greatly stimulates the transfert of cadmium towards the aerial parts at pH 4, which is not the case at pH 6.

2) Amount of phosphorus absorbed

- at pH 4, no significant effect on the phosphorus content of the roots and the aerial parts is observed, irrespective of the treatment;

- at pH 6, major differences appear: cadmium inhibits significantly (1%) the phosphorus contents of the roots and the aerial parts; it can even be observed that cadmium inhibits the phosphorus content of the roots to a greater extent when supplied in mineral form than in humate form (significant at 1 %), which is not the case for the aerial parts. The presence of humic acids does not significantly affect the plant's phosphorus content in comparison with the mineral sample.

IV. CONCLUSION

The association of cadmium with humic acids does not appear to modify the toxicity of the metal; the intensity of its toxic effects remains approximately constant in the pH conditions tested, although the cadmium concentration in the aerial parts of the plant decreased appreciably at high pH values. At pH 4, the introduction of humic acid into the medium facilitates cadmium transfer to the aerial parts, which is not proven at pH 6.

In the absence of cadmium, the introduction of humic acids does not modify the way in which the plant assimilates phosphorus. The latter is inhibited by the addition of cadmium only at pH 6 and, in particular, in the absence of humic acids.

BIBLIOGRAPHY

1. A. GOMEZ, Comparaison de deux modes de préparation de composés humiques associés à différents métaux lourds à pH 6 (en cours de publication).

2. JUSTE C., DELAS, J., 1967 - Influence de l'addition d'aluminium, de fer, de calcium, de magnésium ou de cuivre sur la mobilité électrophorétique, le spectre d'absorption infra-rouge et la solubilité d'un composé humique. Ann. agron., 18, 403-427.

3. JUSTE C., DELAS, J., 1969 - Influence du degré de salification d'un composé humique sur son comportement thermique. Ann. agron., 20, 145-159.

TOLERABLE AMOUNTS OF HEAVY METALS IN SOILS AND THEIR ACCUMULATION IN PLANTS

A. KLOKE

Federal Biological Research Centre for Agriculture and
Forestry, Institute for Non-Parasitic Plant Diseases,
Königin-Luise-Str. 19, D-1000 Berlin 33
Federal Republic of Germany

It is the task of agriculture to produce foodstuffs which in amount and quality fit the demands of the consumers and legislation. It can, however, meet these demands only if the offered aids of production do not bring along a threatening accumulation of heavy metals in soil and plants. As plants take up and accumulate increasing amounts of heavy metals with increasing amounts of hazardous elements in the soil, it was necessary to establish tolerable amounts of heavy metals in soil. They are meant as a precaution, for air and water can be cleaned or are cleaned on their own, soil, however, once contaminated with heavy metals, according to our present day technology and knowledge, can not be cleaned.

In order to show what is hazardous and what is safe one has to draw attention to toxicological considerations of the FAO/WHO Expert Committee on Food Additives in 1972. This working group at that time recommended preliminary amounts for the tolerable weekly intake of lead, cadmium, and mercury. According to these recommendations a healthy man of 70 kg body weight may take in 3.5 mg Pb, 0.525 mg Cd, and 0.35 mg Hg per week (4).

These, as internationally accepted amounts, are starting points of consideration and calculation for the upper limits of these elements in individual foodstuffs including plants. The results of these calculations, estimations, and investigations are the "Guidelines '79 for lead, cadmium and mercury in and on foodstuffs"(1).

Table 1: Guidelines '79 for lead, cadmium, and mercury in
foodstuffs of plant origin (1,4).

Foodstuffs	Guidelines '79 (mg/kg fresh weight)		
	Lead	Cadmium	Mercury
Green vegetables	1.2	0.1	
Sprout vegetables	1.2	0.1	
Fruits	0.2	0.1	
Root vegetables	0.5	0.05	no values
Pomacious fruits	0.5	0.05	given
Stone fruits	0.5	0.05	
Berries	0.5	0.05	
Cereals	0.5	0.1	0.03
Potatoes	0.2	0.1	0.02

The Ministry of Nutrition, Agriculture, Environment, and Forestry of Baden-Württemberg (FRG), writes on this subject:"These guidelines are not to be exceeded, if possible. Where the content of the foodstuffs of plant origin is twice as high as the "Guidelines" indicate the products will be rejected during the ordinary course of the official food quality control and on a commercial basis are no longer allowed to be brought into circulation" (2).

The task of food producers, especially those of agriculture, now is to produce food which fits the high standards of the German Federal Health Office and the requirements of the consumers. To enable agriculture, however, to fit these demands it has been and will be necessary to limit the input of heavy metals into the soil and to list the concentration of heavy metals in soil which according to our present knowledge may be tolerated. That means, if these tolerable amounts are not exceeded or even reached, one can usually assume that the heavy metal content of the plants grown on such soils fits the standards established by legislation. These data are summarized in Table 2.

Table 2: Guidelines for tolerable amounts of some elements (6).
*Provided by the government of the FRG as tolerable upper limits (3).

| Element | Total amount of hazardous elements in airdried soil (mg/kg) | | | |
| | Range frequently found | In heavy contaminated soils | Tolerable amounts | |
			Soil	Sewage sludge
As Arsenic	0.1 - 20	8000	20	
B Boron	5 - 20	1000	25	
Be Beryllium	0.1 - 5	2300	10	
Br Bromine	1 - 10	600	10*	
Cd Cadmium	0.01 - 1	200	3*	20*
Co Cobalt	1 - 10	800	50*	
Cr Chromium	2 - 50	20000	100*	1200*
Cu Copper	1 - 20	22000	100	1200
F Fluorine	50 - 200	8000	200	
Ga Gallium	0.1 - 10	300	10*	
Hg Mercury	0.01 - 1	500	2*	25*
Mo Molybdenum	0.2 - 5	200	5*	
Ni Nickel	2 - 50	10000	50*	200*
Pb Lead	0.1 - 20	4000	100	1200
Sb Antimony	0.01 - 0.5	?	5	
Se Selenium	0.01 - 5	1200	10	
Sn Tin	1 - 20	800	50	
Tl Thallium	0.01 - 0.5	40	1	
Ti Titanium	10 - 5000	20000	5000	
U Uranium	0.01 - 1	115	5	
V Vanadium	10 - 100	1000	50*	
Zn Zinc	3 - 50	20000	300*	3000*
Zr Zircon	1 - 300	6000	300	

It seems necessary now to outline the history of the origin of the guidelines for the tolerable contents of certain elements in agricultural soils. In the beginning this job was only a desk study. It has been well known for a long time that by increasing their content in soil the amounts of most elements in plants grown there are also increased. It was well known too, that certain amounts of a given element in food and fodder plants are toxic to man. At normal amounts of the different elements in soil there will be normal concentrations of these elements in plants too and with these man has been living for centuries and centuries. These normal amounts of the different elements in soil, however, had to be established. As normal contents of the soils such amounts of the elements have been considered, which have been found most frequently in the literature (Table 2, first column; compare to (5)).

Quite normal, however, was a wide range of variation which was specific for each element. In eight cases the highest values found in the first column of Table 2 were regarded as the tolerable amount in soil. For other fourteen elements, however, the tolerable amount in soil was fixed at a higher level and only in one case out of twenty three it was decreased. Also for the heavy metal content of sewage sludge which is applied as fertilizer, upper limits have been established (Table 2) and it is now task of present-day research to support these theoretically fixed amounts.

Heavy metals enter soil through different routes and are slowly accumulated there. Routes by which soil can be contaminated with heavy metals are also dust and rainfall. To limit the accumulation of lead and cadmium in soil by these routes, however, today in the FRG there is also a discussion about the input of these elements via dust and rainfall. The fallout is to be limited as followed:
Cd: $7.5 \mu g/m^2/d \approx 27$ g/ha/a; Pb: $500 \mu g/m^2/d \approx 1.8$ kg/ha/a and the concentration in the air should not exceed 40 ng/m^3 for cadmium and $2 \mu g/m^3$ for lead.

While the pollution of air and water is reversible after closing its source, a contamination of soil with heavy metals normally is not reversible. The contamination of soil is never ending and under normal conditions not reversible. The damage caused by the pollution therefore is not reparable at all. The causer of the pollution is present-day society, the bearers of the risk, however, are future generations. This fact obliges present-day society to keep soil clean and to control the input of heavy metals into the soil and to stop it, if necessary.

To what extent the said "Guidelines" of tolerable amounts of certain elements in soil are correct at the moment is subject of many studies of which only one is to be described.

To find out the amounts of heavy metals and of other elements taken up by plants from soil, in 1975 an arrangement of concrete boxes consisting of 112 concrete frame plots, 1 m^2 each, was made. Homogenized Dahlem-Soil in these plots was supplied with the elements e.g. Cd and Hg. Each element was applied in two concentrations so that in all there were three concentrations including the control. Each concentration consists of four replicates, the control, however, consits of twelve replicates.

In the arrangement of concrete boxes, 1975/76 ryegrass, wheat, and oat were planted successively as precultures and soon after sprouting were incorporated into soil. In June 1976 dwarf beans were planted, harvested, and analyzed (7). In 1976/77 winter rye was planted, in 1977/78 winter wheat, in 1978/79 winter barley, in 1978/79 winter rape, and in spring 1981 potatoes and tomatoes.

It has already been mentioned that with increasing amounts of heavy metals in soil the amount of these elements in plants normally increases too. In Table 3 the amount of mercury in harvested plant material of so far of five cropping years is presented. It is shown that even at 51 ppm Hg in soil the amount of mercury in the grains is below the tolerance level established by the German Federal Health Office.

Table 3: Mercury content of soil and plants after application of $HgCl_2$ to soil in 1975.

Hg applied (mg/kg soil)	0	50	200
Hg found in 1977 (mg/kg soil)	0.18	51	117

Plants		Hg in plant material (mg/kg dry matter)		
Beans	1976 Leaves	0.11	2.1	20.0
	Green bean pods	0.012	0.103	no harvest
Rye	1977 Straw	0.05	0.53	3.45
	Grains	< 0.01	0.03	0.17
Wheat	1978 Straw	0.05	1.13	4.60
	Grains	< 0.01	< 0.01	0.03
Barley	1979 Straw	0.07	0.66	3.33
	Grains	0.01	0.01	0.05
Rape	1980 Leaves	0.10	0.42	3.03
	Grains	0.01	0.01	0.03

In Table 4   analytical results of the cadmium plots are presented. It can be seen that the cadmium content of the soil used in this experiment was at least 8 ppm which was not obvious when the

Table 4: Cadmium content of soil and plants after application of $CdCl_2 \cdot H_2O$ to soil in 1975.

Cd applied (mg/kg soil)	0	50	200
Cd found in 1977 (mg/kg soil) (soluble in 0.5 N HCl)	8	63	300

Plants		Cd in plant material (mg/kg dry matter)		
Beans	1976	Leaves	2.6	4.5	13.1
		Green bean pods	0.5	2.5	1.9
Rye	1977	Straw	0.5	5.3	10.2
		Grains	0.5	0.6	1.9
Wheat	1978	Straw	0.4	3.6	11.2
		Grains	0.5	1.2	2.1
Barley	1979	Straw	1.5	7.0	12.1
		Grains	0.6	1.4	3.5
Rape	1980	Leaves	1.0	29.3	180.0
		Grains	0.2	0.5	7.4

experiment was started. This amount is, of course, already above the tolerable amount of this element in soil which is 3.0 ppm for Cd (Table 2). The cadmium content of the grains in the control, therefore, is already higher than the tolerable amount given by the German Federal Health Office (Table 1).

REFERENCES

1. Anonymous (1979). Richtwerte '79 für Blei, Cadmium und Quecksilber in und auf Lebensmitteln.- Bundesgesundheitsblatt 22,
   282-283.
2. Anonymous (1980). Erlaß des Ministeriums für Ernährung, Landwirtschaft, Umwelt und Forsten des Landes Baden-Württemberg
   über die Schwermetallbelastung von Böden.- GABI 1980,Ausg.B,
   28. Jg., Nr. 39, v. 9.12.1980, S. 1186-1189.
3. Anonymous (1981). Verordnung über das Aufbringen von Klär-
   schlamm (AbfKlär V). Entwurf vom 17.3.1981 - U II 5-530
   115/2.
4. Käferstein, F.K. et al. (1979). Blei, Cadmium und Quecksilber
   in und auf Lebensmitteln.- ZEBS-Berichte 1/1979. Dietrich
   Reimer-Verlag, Berlin.
5. Kloke, A. (1977). Orientierungsdaten für tolerierbare Gesamtgehalte einiger Elemente in Kulturböden.- Mitt. VDLUFA,H.2,
   32-38.
6. Kloke, A. (1980). Richtwerte '80: Orientierungsdaten für tolerierbare Gesamtgehalte einiger Elemente in Kulturböden.-
   Mitt. VDLUFA, H.1-3, 9-11.
7. Kloke, A. (1981). Die Anreicherung von Schadstoffen im Getrei-
   de.- In: Forschungsbericht:"Rückstände in Getreide- und Ge-
   treideprodukten".Hrsg.:DFG. Harald-Bold-Verlag,Boppard,S.30-41.

PLANT UPTAKE OF CADMIUM

SUMMARY OF SWEDISH INVESTIGATIONS UP TO AND INCLUDING 1981

Sven BERGLUND

The National Swedish Environment Protection Board

Health effects

"Cadmium is an environmental, toxic agent that has some characteristics necessitating an extensive study of its occurrence
in order to prevent unfavourable health effects in exposed populations. Friberg et al calculated the average daily cadmium intake
via food necessary to produce kidney damage in a substantial part
of a population to be 250 μg to 350 μg (assuming 4.5 % retention
of ingested Cd, one third of body burden in the kidneys, 50 %
higher Cd concentration in kidney cortex than in whole kidney,
adult kidney weight 300 gm, and biological half-time of cadmium
in kidney cortex some 20 to 40 years). (These assumptions are based
on a review of pertinent scientific evidence.) A joint FAO/WHO expert committee on food contaminants proposed a tolerable daily intake of about 70 μg/day.

The average daily intake in industrialized countries in recent
years has been calculated to be about 40 μg to 60 μg. This means
that the "safety factor" for cadmium intake via food is probably
< 10 μg. The additional intake from ambient air by the general
population is usually negligible, whereas smoking 20 cigarettes
per day may cause an uptake of about 2 μg to 4 μg Cd/day. Assuming
25 to 50 % pulmonary adsorption, this corresponds to a daily intake
via food of 10 μg to 40 μg of cadmium."

From Kjellström, T., Lind, B., Linnman, L., Elinder, C-G.
- Variation of cadmium concentration in Swedish wheat and barley,
Arch Environ Health, Vol 30, July 1975. (11)

Table 1. Cd content in some vegetables and other food stuffs in ug per kg wet weight (from Koivistoinen 1980)

carrot	30	red beet	30
celery root	30	radish	20
potato, new	30	potato	10
turnip	10	cauliflower	10
lettuce	50	spinach	150
onion, yellow	30	pea	30
tomatoe	10	cucumber	5
pear	10	apple	< 2
grape	< 2	orange	< 2
lingonberry	3	spring wheat	40
winter wheat	60	rye	12
barley	20	oats	40

The Cd content in some products are:

wheat bread	30	rye crisp	20
rye bread, sour	20	oat meal	50
margarine	20		

As a comparison the Cd content in some meat and fish products are:

cow beef	< 5	pork butt	5
horse meat	100	liver (cow)	120
liver, steer	30	liver, pig	70
kidney, steer	200	kidney, pig	180
whole milk	< 2	chicken	< 5
liver paste	20	cod	< 5
perch	< 5	baltic herring	5
salmon	5	eel	50
mussels in water	240		

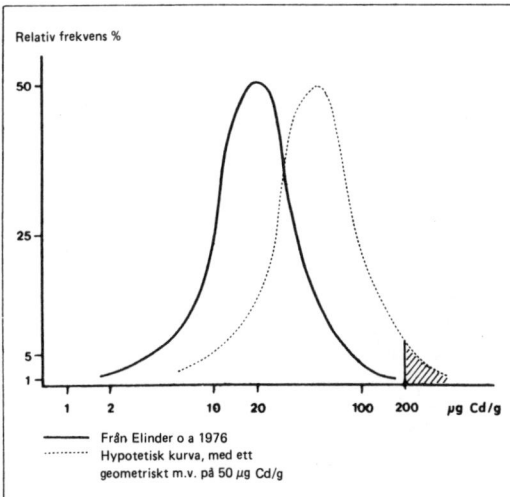

Figure 1 : Distribution of frequence over Cd concentrations in kidney cortex of Swedes 1974 (left curve) and corresponding distribution at a geometric mean of 50 ug Cd/g (right curve) (9)

The distribution of the amount of Cd in a population follows
the pattern in Figure 1. The left curve is the distribution in
kidney cortex found in 1974 in Sweden with the top of the curve
near 20 µg Cd/g in kidney cortex. If the daily intake increases
resulting in the geometric mean value of 50 µg, the curve is
moved to the dotted position, a certain number (about 5 %) of
the inhabitants will exceed the limit 200 µg Cd/day (which is
the risk limit for chronic kidney damage). (Elinder et al 1978).
- Studies in Sweden of horse kidneys show that changes in the
kidneys start to develop already at 50-70 µg Cd/g. WHO reported
in 1976 an increased risk for prostata cancer at a certain Cd
exposure.

The main source of the human Cd uptake is the diet. As a certain
portion of the Cd in the soil-plant system today originates from
air deposition, the Cd content in the diet will vary with the air
deposition as well as with other factors.

Human food intake

Interesting information was given 1980 by Koivistoinen (12) regarding mineral element composition of Finnish food. Out of the daily
intake of 13 µg the Cd food sources were:

cereals	36 %
meat	9 %
fish	3 %
dairy products	12 %
vegetables	29 %
other	8 %

A Swedish investigation (10) about levels of lead, cadmium and
zinc in vegetables shows that the daily intake from vegetables
is 5 µg Cd including root vegetables and potatoes.

The main food sources are thus cereals and vegetable foods. See
Table 1. Even meat and dairy products have their origin in plant
production. The circulation of Cd through farm land is apparently
of great importance.

High levels of Cd is found even in liver and kidney from moose
(Alces alces), see Table 2 (14).

Table 2. Cadmium concentration in liver and kidney from
moose shot in 1980 (mg/kg wet weight) (14)

Age, years	Liver Quantity	Mean	Range	Kidney Quantity	Mean	Range
< 1	8	0.15	0.07-0.23	16	0.45	0.18-0.81
1.5	20	0.31	0.07-0.53	18	1.6	0.64-3.4
> 2	41	0.58	0.10-2.4	35	3.5	0.88-13

The wide distribution of cadmium concentration in Table 2 indicates
that other factors besides age influence the uptake. The high levels
of Cd were found in the south of Sweden. One explanation for this
is the higher atmospheric deposition in this area. Wild grazing
animals are highly dependent both on the foliage and surface water.
The Cd levels in liver and kidneys from cows in the same area were
about one fourth of those from moose at corresponding age. Kidney
from moose has been banned for consumption (except kidneys from
yearlings, which can be consumed one to two times per month).

Cadmium in soils

The Cd content in Swedish soils is reported by Andersson, (19)
1977, see Table 3. The mean content was 0.22 mg/kg for both cultivated soils (186 samples) and non-cultivated soils (175 samples).
This result gave the calculated content of Cd at 20 cm depth of
550 g/ha for both groups of soils. The great variation of the content according to Table 3 should be observed. The reason for this
can be both natural and anthropogenic. The cultivation of the soil
including fertilization has apparently not affected the average Cd
content.

Table 3   Cadmium content of Swedish soils. After A Andersson:
           Heavy metals in Swedish soils. On their retention,
           distribution and amounts. Swedish J. Agric. Res. 7:7-20,
           1977.

Number of samples: cultivated   186 ⎫
                     not cultivated 175 ⎭ 361

| Content | | |
mg/kg (PPM)	g/ha (0-20 cm)	% of soils in each group
0,063	158	4
0,063-0,124	158- 312	18
0,125-0,249	313- 624	52
0,25 -0,49	625-1 249	21
0,50- 0,99	1 250-2 499	4
0,99	2 499	1

Mean	0,22	550
Cultivated soils	0,22	550
Not cultivated soils	0,22	550

The supply of metals and removal with harvest is described generally in Table 4.

Metal	Soil content	Supply			Removal with harvest		
		Air depos.	Fertil.	Sludge	Wheat	Barley	Pota-toes
Cadmium	550	2	2	10	0,3	0,1	1
Cromium	39 000	10	15	100	2	2	6
Coppar	37 000	20	50	500	30	25	40
Mercury	150	0,5	0,02	5	0,1	0,1	0,05
Lead	40 000	100	15	200	3	2	3
Zinc	149 000	150	110	1 500	150	100	100

Table 4.   Heavy metals in Swedish soils. Average figures in g/ha for supply
and removal. Local variations are often great. (15)

The contamination of soils from neighbouring industries will affect
the Cd content, see Figure 2. These local disturbancies are described in both Swedish and foreign papers and will not be further
discussed here.

The long-range transportation of Cd has increased with the use
of the metal Cd and the burning of fossil fuels. Even the use of
zinc for plating has resulted in the diffusion of Cd, as Cd is
an impurity in Zn. Thus the metal Cd is well spread in society
and will be redistributed through incineration, the food chain etc.

The deposition from air will vary with the air streams and dis-
tance from source. Recent reports are reviewed in Table 5. In
addition to this Tjell and Larsen (Denmark) have reported from
measurements 1973-78 decreasing amounts (from 6-7 g/ha   to 2-3

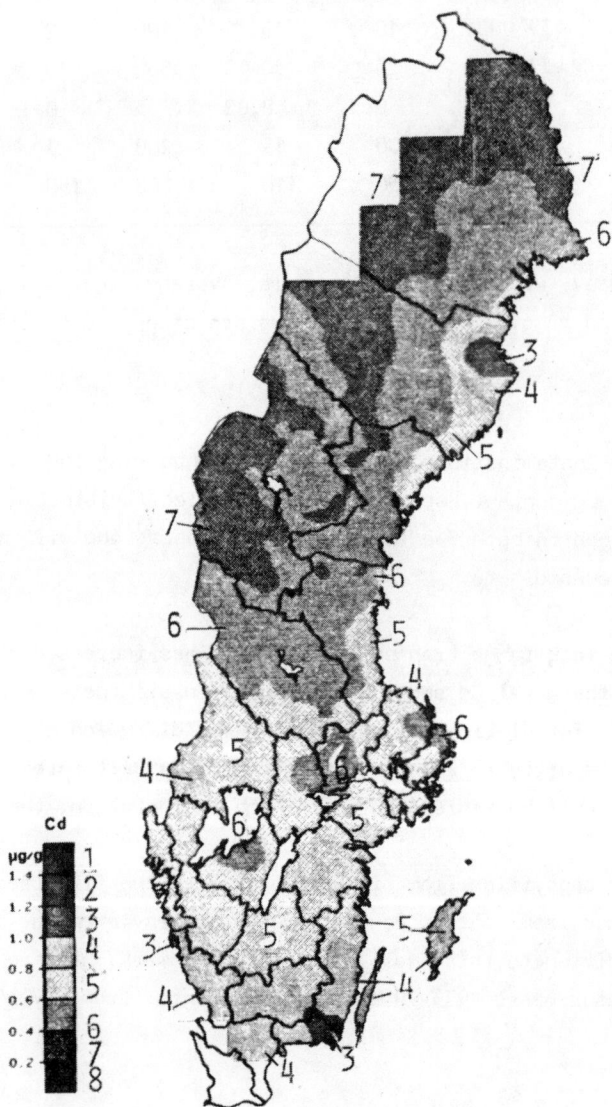

Figure 2. Cadmium content (µg/g d.w.) in moss samples collected in 1975 (16)

Table 5. Atmospheric deposition of Cadmium (SNV 1981)

Local	g/ha, year	Method
Linderödsåsen (Tyler) S Sweden	7	Sphagnum
Smålandsstenar " S Sweden	3,5	"
Ulricehamn " SW Sweden	4,5	"
Forshult (VO-hamn) S Sweden (Rühling)	5,5	Snow
80 km V Skellefteå N Sweden (IVL)	2,0	Snow
Uppsala (A Andersson) E Sweden	1,0	Funnel
Fyresdal S Norway (Maagen/Langeland)	10	Snow prophiles
Danish statistics	1,1-5,0	Funnel
Danish (MF Hovmand)	2,0	

g/ha at the end of the period). Andersson (1) has reported 0.9 g/ha
in 1977 at Uppsala (in the middle of Sweden) and 0.75 g/ha from measurements 1976-78 on precipitation, which excludes dry deposition in both
cases. Including dry deposition the increase of Cd content in soils from
atmospheric deposition has been determined to 0.2 % annually in the Uppsala region. This figure is probably three times higher in southern
Sweden (1) (compare the Danish results).

The loss of Cd with the drainage water is calculated to be 0.023 g/ha
annually (median value) from investigations measuring Cd content in
drainage waters (1).

Plant uptake of cadmium

Andersson (4) has reported from an investigation regarding composted municipal wastes about plant uptake from sludge application
to soil. Both pot trials and field trials are included. Here
only the field trials are reported. The soil had a relatively
high content of humus and clay, which will check the Cd-uptake
in plants. The Cd content in sludge used for application was 4.6-
4.9 $\mu g/g$ dry matter. The corresponding figure for farm yard manure
used in the same investigation was 0.17-0.51 (mean 0.33). The
application rates 50 and 100 tons DM/ha had no influence on the
pH of the soil, which was 6.4. The highest application rate of
100 tons DM/ha gave an increased uptake of Cd in lettuce of 13 %
as a mean of four years.

Trials with moderate application rates of 5 and 10 tons of DM/year
increased the Cd content in the soil with 20 and 41 % during the
four year period. The amount of natural Cd was 460 g/ha down to
20 cm depth.

Compared with the fertilizer application there was no difference
in the uptake of Cd in the grain in barley 1975, less in oats 1976,
no difference in spring wheat 1977 and less in fodderrape in 1978.

Calculation was made on the percentage of Cd removed with harvest
compared with the addition. For the period 1975-78   284 % of
Cd added by fertilizer (NPK 20-5-9) was removed, and 3.1 and 1,7 %
was removed respectively at 5 and 10 tons of sludge application.
As very little is leached out, the remainder is accumulated in
the soil. The uptake of Cd in lettuce varied with the precipitation in the above-mentioned investigation.

The precipitation during the growing season May-August showed a
direct influence on the uptake, see Figure 5. Roughly, high precipitation means high Cd content in the wheat grain (9).

Andersson has also reported on the "Influence of organic fertilizers on the solubility and availability to plants of heavy metals in soils" (3). He states that "from both laboratory and field experiments it can be seen that organic fertilizers added to soils may reduce the solubility and uptake in plants of heavy metals, despite high contents of these elements. This may be the case even if the total content in the soil is considerably increased. Whether the solubility and availability to plants are reduced or not depends on competition between the reducing effect on the solubility from the organic matter of the fertilizer and the opposite effect from the quantity of heavy metals added together with the organic matter. Other important factors are the type of organic matter and the physico-chemical properties of the soil."

The results of applications of 50 and 100 tons of DM of compost and sludge are presented in Table 6. An increase of the cadmium content of 10 and 50 % respectively from sludge treatment has occurred in the case of carrots. Both lead and mercury content, however, have been depressed by sludge treatment.

When phosphorus is added as fertilizer a certain amount of cadmium is even added to the soil.

For barley significant positive correlations were found for grain versus added amounts of phosphorus and different fractions of soil Cd. At an annual addition of 25 kg P/ha and year the   percentage increase of Cd were annually 1,1 % for grain and 0,55 % for straw during 1963 to 1978 (5).

Andersson and Nilsson states (6) from pot trials that single applications of 20 tons dw per ha of sludge can be tolerated on cultivated soils. But repeated application of 7 tons dw per ha gives long-term effects such as increased levels of Zn, Cu, Ni, Cr, Pb, Cd and Hg in the soil and of Zn, Cu, Ni and Cd in plant material.

Differences of Cd uptake between different wheat cultivars has
been as great as 30 % (7). Four different cultivars grown in
five different regions have shown that other factors apart from
genetic have greater influence on the uptake. Higher uptake in
southern Sweden is believed to be caused by higher atmospheric
deposition and higher commercial fertilizer rates. The concentrations of Cd in wheat grain from Skåne (in the south of Sweden)
was found to be 96 ng Cd/g dw as compared to 66 ng Cd/g dw in
wheat from Uppland (in the middle of Sweden). There are similar
differences in soil Cd levels which together with other soil factors may explanin the differences found in the wheat grains.

The way of transportation from the uptake zone in the soil seems
to affect the Cd content in different parts of the plant.

Table 6. Heavy metals in plant material (carrots) for three metals
          from field trial (n = 4), μg/g dw (3).

Treatment	Pb	Cd	Hg
0	$0,69^{\pm}0,26$	$0,09^{\pm}0,02$	$0,09^{\pm}0,05$
Compost			
50 tons dw/ha	$0,38^{\pm}0,08$	$0,10^{\pm}0,01$	$0,08^{\pm}0,03$
100 tons dw/ha	$0,25^{\pm}0,03$	$0,12^{\pm}0,01$	$0,05^{\pm}0,05$
Sewage sludge			
50 tons dw/ha	$0,27^{\pm}0,08$	$0,10^{\pm}0,00$	$0,02^{\pm}0,00$
100 tons dw/ha	$0,20^{\pm}0,03$	$0,14^{\pm}0,01$	$0,02^{\pm}0,00$

The availability for plants regarding the three toxic metals lead, mercury and cadmium differ in such a way that the uptake of Pb and Hg in plants is very little affected by the concentration in the soil while the uptake of Cd is higher at higher concentrations in soil (Pettersson 1979). Cd is bound looser in the soil compared with other metals. As the Cd content in the soil naturally is low (0.55 kg/ha for Cd as a mean compared with 40 kg/ha for Pb) even limited additions of Cd will affect the content of the soil.

Stenström and Lönsjö have used two different radioactive isotopes of cadmium to label sludge in a micro-plot technique (17). On a subsoil of sand 32 plots were covered with seven different topsoils to a depth of 20 cm. The application rates were 25 tons/ha of dry matter as sludge at one treatment and 5 tons/ha annually. The up-take of Cd after 25 tons application was about the double amount for silt loam and (loamy) sand compared with loam after the first year. The higher application rate gives a higher Cd content in the wheat grain the first year. The soil factors however has a certain influence on the total Cd content. Lönsjö has reported 1980 from the same investigation (8) that a deposition of 1 g Cd/ha and year gives an additon to the wheat grain corresponding to 0.1 μg per kg dry matter as a mean value. If the Cd content in the wheat grain is 50 μg per kg DM the annual deposition on the soil thus gives an increase in the grain of 0.2 % of the deposition.

The conclusions of the 1972-1979 trial period are: (see Figure 3):

- higher uptake of Cd was found with clay loam, clay and sandy clay than with loam, silt loam and (loamy) sand, which is ex-plained by a higher content of organic matter for the lighter soils,

- the addition of lime (10 tons $CaCO_3$/ha) has depressed the Cd content in the grain to about the half for the clay loam but only with 10 % for the silt loam, in spite of a pH increase of one unit, which in this case is partially explained by a low buffert capacity in the soil,

- 0.3-0.6 % of a cadmium addition to the soil is removed by
  the harvest of the grain from spring wheat,

- low pH values in soils gave high uptake of Cd in the grain,

- the application of 25 tons of DM 1972 as sludge has not increased the Cd content in the grain during 1973-79 at the clay loam
  and has decreased the content with about 15 % at the silt loam,

- the total uptake of Cd showed no significant difference when
  one 25-ton application of sludge was compared with five 5-ton
  applications (except for the first year when the soil structure
  favoured uptake of all elements),

- the uptake of Cd was found to decrease in the order wheat grain >
  > red clove > oat grain > barley grain.

The influence of pH on Cd uptake has been described by several
authors. Andersson and Nilsson show this clearly in Figure 4.

Cadmium uptake even takes place directly from the atmosphere.
Danish investigations show that English ryegrass takes 30-40 %
of its Cd from the atmosphere. This may explain high Cd contents
in leafy vegetables. Dust from certain industries may increase the
intake still more.

The uptake of Cd in different harvest products with different
treatments is demonstrated in a Danisch investigation, see Table 7.
(8). 30 ton DM as sludge with low Cd content (3 mg Cd/kg DM) gives
no or little effect compared with commercial fertilization for barley grain, Italian ryegrass, sugar beets, cabbage and potatoes
while higher contents were found in carrots and spinach. A 30-ton
application of sludge with high Cd content (77 mg Cd/kg DM), however, gave still no increase in barley grain and cabbage but 70-
100 % increase for Italian ryegrass, sugar beets, potatoes and
carrots. Some of these results may be explained by soil factors
as pH and cation exchange capacity.

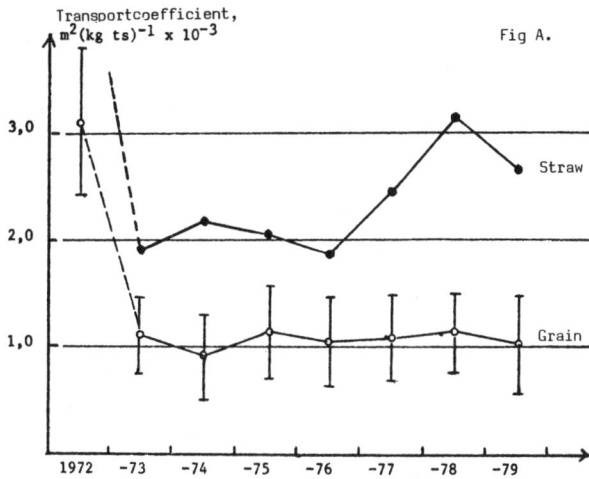

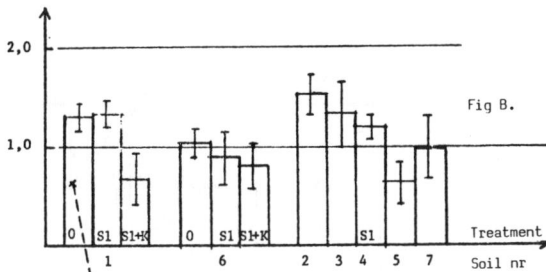

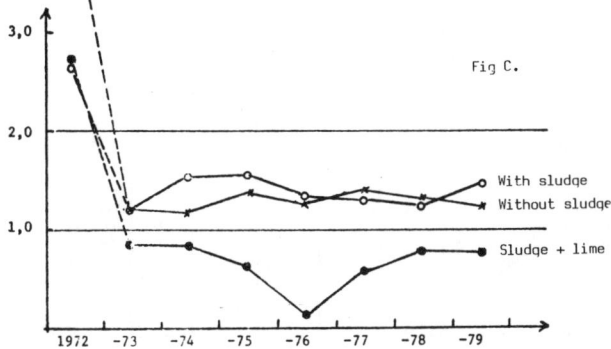

Figure 3: Uptake of Cd in springwheat (8)

      A. Mean values of soils and treatments

      B. Three treatments:

         O = without sludge, S1 sludge treatment 25 tons
         per ha and year, K = liming 10 tons $CaCO_3$ per ha.

      C. Effect of liming. Transportcoefficient is
         defined:

$$\frac{\text{unit Cd (kg ts)}^{-1}}{\text{unit Cd m}^{-2}} = m^2 \, (\text{kg ts})^{-1}$$

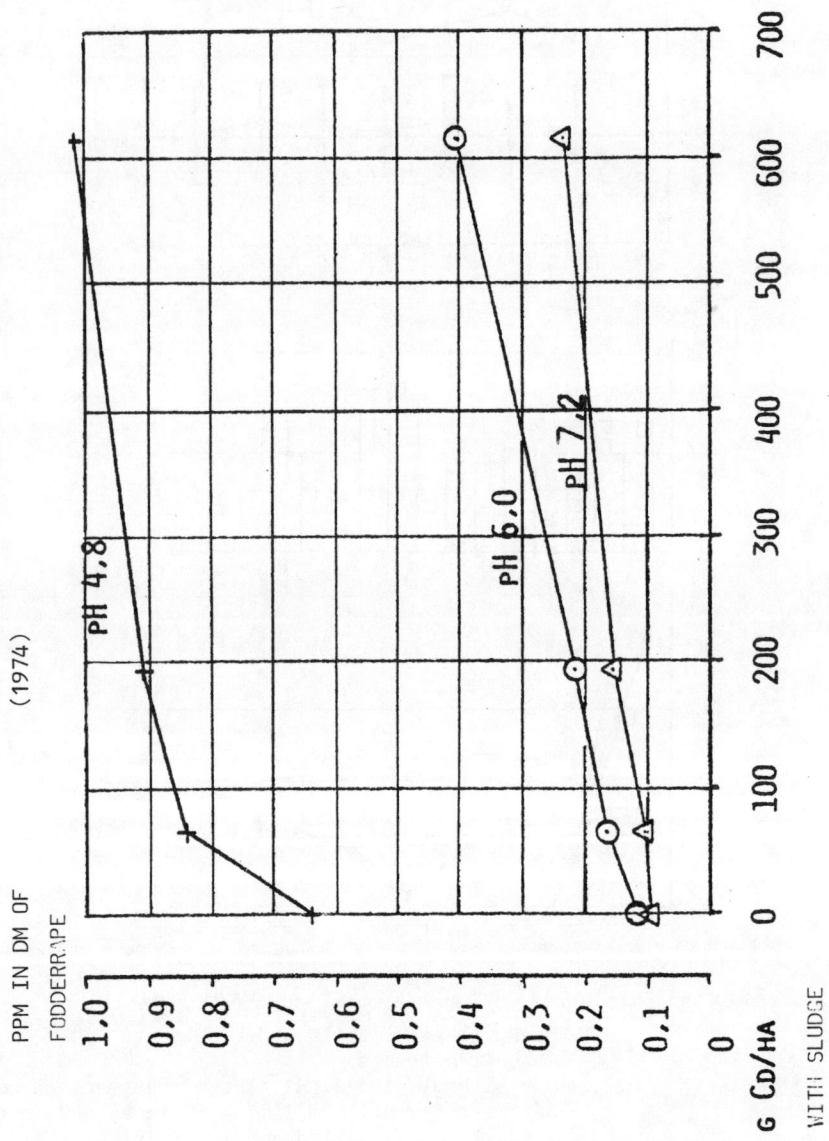

FIGURE 4.   The influence of pH on the uptake cadmium by rape
After A Andersson and K O Nilsson: Influence of lime and
soil pH on Cd availability to plants. AMBIO. Vol.3 no 3
(1974)

Table 7. Cadmium content of some plant products at different fertilizing (Danish survey) (8).

Plant	n	mg Cd per kg DM (PPM)			
		Without fertilizer	Fertilizer 150 N+62 P [1]	30 tons sludge TS/ha	
				low Cd content [2]	high Cd content [3]
		Cd- supply g per ha			
		0	6	90	2 300
Barley, seed	5	0.04	0,06	0,07	0,07
Barley, straw	5	0,15	0,17	0,17	0,18
Italian ryegrass	5	0,12	0,17	0,16	0,38
Sugarbeets roots	5	0,16	0,26	0,25	0,55
Sugarbeets tops	5	0,50	0,61	0,55	1,26
Potatoes	5	0,17	0,20	0.15	0,33
Carrots	4	0,53	0,62	1,09	1,06
Greenkale	3	0,11	0,13	0,12	0,14
Cabbage	1	0,65	0,63	0,65	0,63
Spinach	2	1,60	1,79	1,90	5,92 ?

1) Cd content estimated to ca 100 mg per kg P

2) 3 mg Cd/kg DM

3) 77 mg Cd/kg DM

The addition of sludge not only increases the amount of certain elements but even changes the physical and biological environment in the soil.

The direct deposition from the atmosphere is of little importance for Cd in small grain which is sheltered by the husks. This was indicated when a comparison was made with lead uptake, which to 90 % or more is taken up directly from the atmosphere. On the contrary leafy material will take up Cd from the atmosphere.

The Cd content in agricultural soils in central Sweden is around 75 % of those in sothern Sweden, which explains some of the differences in the plant uptake in wheat grain. The same differences is found for Zn in grain.

The additon of total 105 g Cd per ha during 23 years with phosphorus fertilization (annually 45 kg P per ha) has increased the Cd-content in winterwheat with 20-25 % (S L Jansson 1980) (8). Jansson also stresses the importance of the analyzing methods and the risk for contamination of the samples during handling and preparation.

Percolation trials with columns of different soils have shown that the mobility increased in the soil in the order Cr < Pb < Hg < Zn < Cd < Ni. Coarse texture and acid soils retained Ni, Cd and Zn relatively poorly (Lotse and Preston 1980) (8).

Numerous investigations regarding plant uptake of radioactive isotopes from soil and atmosphere in Sweden have shown that especially fodder species often have a much more effective uptake of these elements from the atmosphere than from the soil. The uptake from atmosphere is, however, very much dependent on the climate, especially the precipitation. A small rain means a more efficient uptake than a heavy rain. A heavy rain menas washing off (8).

Nilsson has reported (18) from a field trial started 1956 about
the conversion of organic matter etc when using different nitrogen fertilizers and organic materials. The sludge plots have
received 7 tons of dry matter or 188 kg N/ha as total N as a mean
until 1975 compared with 80 kg total N for manure and 80 kg N/ha
for nitrate of lime (calcium nitrate). The difference in nitrogen
application is reflected with a higher level of harvest as dry
matter, see Figure 5. The carbon content in the soil has increased for sludge treatment, but the pH has decreased. Up til now this
development of the pH seems to be the most negative result of this
trial regarding sludge application. During the 19 year period the
harvest level, however, has increased with 27 % for the sludge plots.

The Cd content in the sludge was not measured until 1969. The content values lies between 4.6 and 13.2 ppm with the lowest figure
1975. The Cd content in the grain from sludge treatment compared
with manure treatment was the following:

| | | Cd content ug/g at treatment | |
Year	Harvest	Sludge	Manure
1971	Barley	0,067	0,053
1972	Oats	0,070	0,024
1973	Barley	0,067	0,031
1974	Fodderrape	0,303	0,165 (green mass)

A higher content of Cd with sludge application can be noted.
The same was recorded regarding Cd content in the straw.

Figure 5. Harvest development as dry matter harvest in percent of the years mean harvest for the total trial. K O Nilsson 1980.(18)

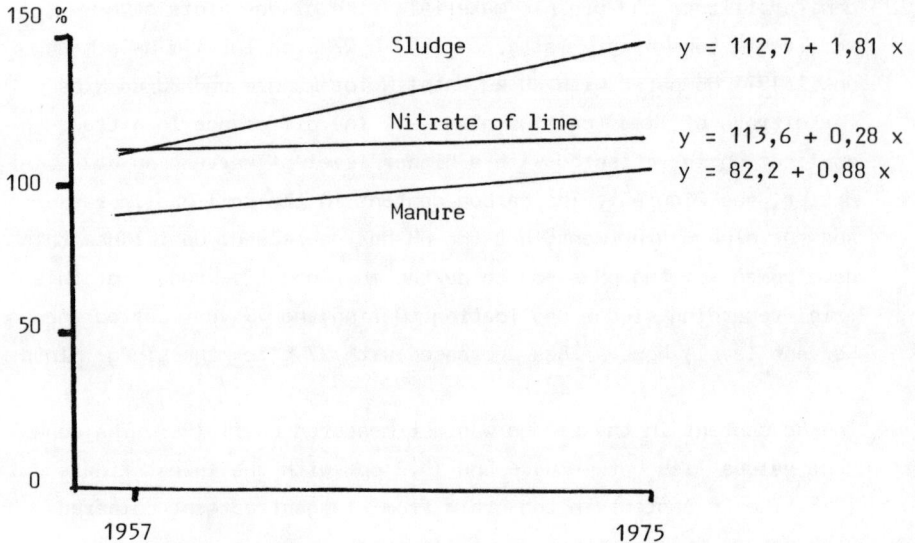

$y = 112,7 + 1,81 \ x$ (Sludge)

$y = 113,6 + 0,28 \ x$ (Nitrate of lime)

$y = 82,2 + 0,88 \ x$ (Manure)

Application kg/ha x year	Development in the soil		
Total N		Carbon,%	pH
	Initial 1957	1,50	6,54
	Situation 1975		
188	Sludge	2,23	5,99
80	Nitrate of lime	1,38	6,68
82	Manure	1,89	6,64

Andersson (2) has given a Cd-balance for Swedish cultivated soils 1977. This has been modified in 1982 (personal communication):

g/ha x year

Supply

Fertilizers and lime	2.0
Precipitation	0.7
Total supply	2.7

Removal	0.8
Leaching	0.1
Total removal	0.9

Supplied in excess is 2.7-0.9 = 1.8 g/ha x year.
The annual increase of the Cd content in cultivated soils is thus 0.3-0.4 % for central Sweden. The atmospheric addition is probably close to 2 g/ha in south Sweden which speeds up the annual increase in the soils to 0.5-0.6 %.

Table 8. Heavy metals in sewage sludge from 114 sewage works analysed 1980-81 at SLL, Sweden. Mean values, mg/kg DM compared with "normal contents" from 1968-71. The three sizes of sewage works represent 11, 33 and 42 % of the total connected. (SLL 1981)

Metal	Number of connected inhabitants (80-81)			"Normal contents" 1968-1971	No of works over "normal contents" -81
	5000	5-20000	20000		
Hg	2	4	4	4-8	2
Cd	2	3	5	5-15	2
Pb	73	96	132	100-300	1
Cr	69	82	344	50-200	10
Co	4	6	7	8-20	1
Ni	14	28	40	25-100	3
Cu	340	363	452	500-1500	1
Zn	700	1007	1454	1000-3000	2

The development of heavy metals in Swedish sludge can be described with the statistics in Table 8.

The sewage works tested 1980-81 are not the same as those tested 1968-71 why the figures are not completely comparable. Both investigations are reasonably representative, however.

From Table 8 it can be seen that the heavy metal content has decreased during the 10-year period. For cromium a new analysing method has increased the values with 60 % compared with the older method.

The pressure from authorities to cut down on metal pollution together with improved processing techniques will explain the decreasing figures of metals in sewage sludge.

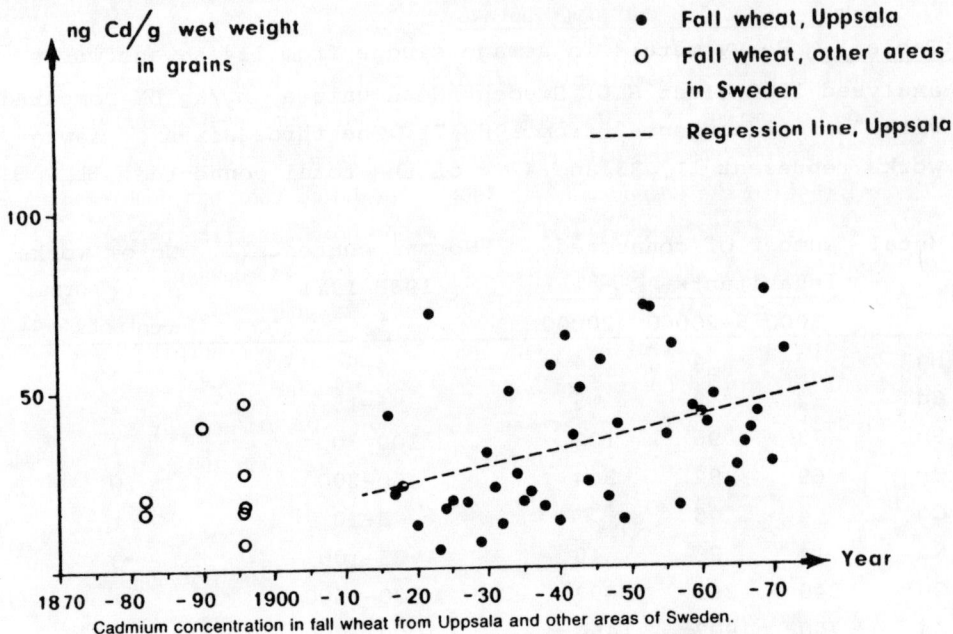

Figure 6 - Cadmium concentration in fall wheat from Uppsala and other areas of Sweden.

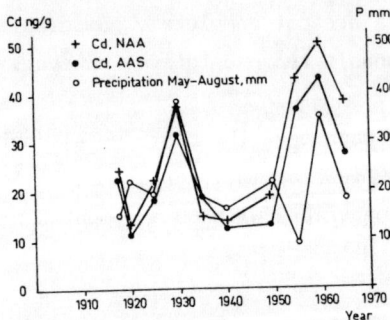

Figure 7 - Precipitation during the growing season May-August and Cd-concentrations in grain of winter wheat, analysed by means of neutron activation (NAA) and atomic absorption (AAS). Wheat analyses from Kjellström et al. (1975), precipitation data from the meteorological station at Ultuna (Rodskjer, 1980).

Long term development of Cd uptake

Kjellström et al have tested archive samples of winter wheat regarding Cd content (11). Despite the large individual scatter among the samples from the different years, Figure 6, a significant increase in the Cd content was found. The Cd content has doubled from 1916 to 1972. Kjellström et al found no covariation between Cd levels and climated factors during this period. However, if precipitation data for the growing season May-August was regarded, certain similarities was found, see Figure 7. High precipitation resulted in high Cd contents. If one special year (1955) was excluded, the correlation becomes significant. The Cd content of the wheat grain increases approximately 10 ug/g for each 100 mm of precipitation.

The long term increase of Cd in winter wheat is probably caused by the diffusion or dissipation of Cd over the world described at page 4 above. The variation of climate factors will apparently cause a great variation of uptake in plant material.

Conclusions

From the investigations referred to in this paper it can be concluded that
- The uptake of Cd in agricultural plants is dependent on:
  precipitation during growing season,
  atmospheric transportation of Cd,
  genetic differences of plant material including different cultivars within the same species
  many different soil factors (pH, organic matter content, cation exchange capacity etc)
  Cd content in the soil

- A big portion of Cd can be taken up directly through the leaves from the atmosphere

- Continuous sludge application with moderate rates has improved
  carbon content (humus factor) and soil structure but decreased
  pH and caused increased Cd content in plant material

- Single moderate applications of sludge will at normal conditions
  cause only slight changes of the Cd content in small grain

- Food habits will affect the cadmium intake

- The long term effects of rising cadmium content in the vegetable
  food products will gradually reach a toxic level for certain
  groups of people.

REFERENCES

1. Andersson A. Cadmium and lead in precipitation and drainage water. Deposition and leaching in the Uppsala area. Swedish J. Agric. Res. 11:119-125, 1981.

2. Andersson A. Heavy metals in commercial fertilizers, manure and lime. Cadmium balance for cultivated soils. Reports of the Agricultural college of Sweden. Seria A, Nr 283, Uppsala 1977.

3. Andersson A. Influence of organic fertilizers on the solubility and availability to plants of heavy metals in soils. Grundförbättring, 27, 1975/76:4, 159-164.

4. Andersson A. Results from fertilization experiments with composted municipal refuse. SNV PM 1097, Swedish Environment Protection Board 1979. (Summary in English)

5. Andersson A and Hahlin M. Cadmium effects from phosphorus fertilazation. Swedish J. Agric. Res. 11:3-10, 1981.

6. Andersson A and Nilsson K O. Influence on the levels of heavy metals in soil and plant from sewage sludge used as fertilizer. Swedish J. Agric. Res. 6:151-159, 1976.

7. Andersson A and Pettersson O. Cadmium in Swedish winter wheat. Swedish J. Agric. Res. 11:49-55, 1981.

8. Cadmium in the soil-plant environment. Lectures and discussions at a conference at the Royal Swedish Academy of Agriculture and Forestry on June 5th, 1980. Rapport nr 4, Stockholm 1980. (Short summary in English)

   - Gunnarsson O. Cadmium in the soil-plant environment, p 4-47

   - Andersson A and Pettersson O. The content of cadmium in winter wheat p 48-61

- Jansson S L. The annual variation of Cd in crops, p 62-69

- Lotse E and Preston E E. The mobility of cadmium and some other heavy metals in soils, p 70-85

- Lönsjö H. The plant availability of cadmium under field conditions. Some results from field trials of several years duration with radioactive isotopes, p 86-97

- Nygård B and Stålberg S. The cadmium issue from the viewpoint of agricultural chemistry, p 98-100

- Dellien I. Remarks on the determination of cadmium in fertilizers, soils and plants, p 101-107

- Umebayashi M. Some problems on cadmium soils and plants in Japan, p 108-113

9. Elinder C-G, Friberg L and Piscator M. Harmful effects of cadmium. Läkartidningen 75:4365-4368, 1978. (Short English summary)

10. Fuchs G, Haegglund J and Jorhem L. The levels of lead, cadmium and zinc in vegetables. Var föda 28:160-167, 1976. (Short English summary).

11. Kjellström T, Lind B, Linnman L and Elinder C-G. Variation of cadmium concentration in Swedish wheat and barley. Arch. Environ. Health / Vol 30, 321-328, July 1975.

12. Koivistoinen P. Mineral element composition of Finnish foods: N, K, Ca, Mg, P, S, Fe, Cu, Mn, Zn, Mo, Co, Ni, Cr, F, Se, Si, Rb, Al, B, Br, Hg, As, Cd, Pb and Ash. Acta Agriculturae Scandinavia, Supplementum 22, Stockholm 1980.

13. Lenander E. Tungmetaller i avloppsslam - begränsande faktor vid slamanvändning inom jordbruket. Umeå universitet, Rapport F 10, 1981. (Swedish only)

14. Mattsson P, Albanus L and Frank A. Cadmium and some other elements in liver and kidney from moose Alces alces). Vår föda 33:335-345, 1981, (Short English summary)

15. Pettersson O and Ericsson J. Tungmetaller och avloppsslam i jordbruket. Aktuellt från lantbruksuniversitetet 274, Uppsala 1979. (Swedish only)

16. Rühling Å and Skörby L. Landsomfattande kartering av regionala tungmetaller i mossa. SNV PM 1191, 1979. (Out of print)

17. Stenström T and Lönsjö H. Cadmium availability to wheat: A study with radioactive tracers under field conditions. Ambio Vol 3 No 2 87-90, 1974.

18. Nilsson K O. Development in harvest and conversion of organic matter when using different nitrogen fertilizers and organic materials. Studies in a small-plot field trial during 20 years. Swedish University of Agricultural Sciences, Department of Soil Sciences. Report 127, Uppsala 1980.

19. Andersson A. Heavy metals in Swedish soils. On their retention, distribution and amounts. Swedish J. Agric. Res. 7:7-20, 1977.

CADMIUM SPECIATION IN SOIL SOLUTIONS OF SEWAGE SLUDGE AMENDED SOILS

A.R. TILLS and B.J. ALLOWAY

Department of Botany and Biochemistry
Westfield College, University of London

Summary

The soil solution is the major source of metals taken up by plants
from soils.   Samples of soil solution were fractionated by HPLC gel permeation columns to investigate the main cadmium containing species present.   It
was found that all the cadmium occurred in a UV absorbance peak associated
with low molecular weight organic molecules which would also contain inorganic compounds and Cd^{2+} ions.   Further studies with a narrower separation
range column in the low molecular weight region revealed that all the cadmium occurred in the largest UV absorbance peak.   These low molecular
weight cadmium species may be ionic, inorganic complexes or small organic
complexes such as organic acids.   It is intended to use reverse phase
chromatography to separate these.

The results obtained so far and earlier work with other techniques
support the conclusions of other workers using computer model predictions,
that most of the cadmium in the soil solution of sludged soils is cationic.

1. INTRODUCTION

Cadmium, unlike many heavy metals, is comparatively labile in the soil-plant system and is therefore more readily taken up by plants. With the increasing concern over its effects on human health a great deal of work has been carried out in recent years on the soil-plant relationships of cadmium. Consequently, much is now known about this element, especially its adsorption by a wide range of soil types and the variations in its uptake between species and cultivars of crop plants.[1] However, there is still a lack of information about the forms of cadmium (and many other metals) present in the soil solution. This is the major source of all elements taken up by plants and therefore the speciation of metals present in this solution is likely to have a marked effect on plant uptake. However, microorganisms and root exudates in the rhizosphere may modify the soil solution constituents reaching the roots by mass flow and diffusion.

Unfortunately, studies on soil solutions are rendered difficult by the very low concentrations of metals and molecules present in them.

One approach to this problem has been the use of computer models such as "GEOCHEM", to predict the metal species likely to occur in soil solutions, under given conditions, on the basis of known chemical equilibria and properties of the soils concerned.(2,3 and 4) This exciting development has great potential in soil-plant relationship studies but is dependent upon the reliability of the data incorporated into the model. There is a continuing need for independent detailed analytical studies of metal species in soil solutions and their plant availability to assess the accuracy of the predictions and also to provide more detailed information to improve the "fine tuning" of the model.

Research into the speciation of cadmium in soil solutions has been in progress at Westfield College since late 1978, funded by the Department of the Environment. During this time, the most rewarding line of investigation has proved to be the fractionation of soil solution consituents by liquid chromatography and their subsequent analysis for cadmium by flameless atomic absorption spectrophotometry.

Several chromatographic techniques have been tried but not all were suitable for use with very dilute soil solution. Gel permeation chromatography using Sephadex^R gels gave only mediocre resolution and were very time consuming. The development of High Performance Liquid Chromatography (H.P.L.C.) over the last ten years has enabled much improved resolution to be achieved in gel permeation chromatography. This is due to the use of high efficiency columns packed with small particle supports (<10μm) which

has led to a narrowing of the chromatographic zones and consequent increased resolution.   Most of the results reported in this paper were
obtained using HPLC.

2. METHODS

A range of trace metal-polluted soils was collected from sites in
Britain and some other countries.   The major sources of contamination
were sewage sludges and metalliferous mining.(5)

Prior to the extraction of samples of soil solution, all soil samples
were equilibrated with deionised water at field capacity moisture status
for at least two weeks.   Samples of these moist soils were centrifuged at
3500rpm for 30 minutes in modified centrifuge tubes.

The soil solution obtained was passed through a 0.45µm membrane filter
and aliquots were removed for total metal analysis and pH measurement.

The sample was injected (Waters U6K injector) into an HPLC system
fitted with either a µ Bondagel E-500 or a µ Porasil GPC 60Å column (Waters
Associates, Milford, Massachusetts).   Both columns were 300mm x 4mm internal diameter.

Nominal Molecular Weight Separation Ranges of Columns Used

Column	Lowest Molecular Weight	Highest Molecular Weight
µ Bondagel E-500	5,000	500,000
µ Porasil GPC 60Å	100	10,000

A mobile phase of 100% water was used with the µ Bondagel column and
50% methanol in water with the µ Porasil column.   A flow rate of 0.5ml
min^{-1} was supplied by a Waters 6000A pump.   The UV absorbance of the
column eluate was measured at 254nm and fractions collected and analysed
for cadmium using a graphite furnace atomic absorption spectrophotometer
with automatic background correction.

The void volumes of the columns (Vo) were found using Blue Dextran
2000 (Pharmacia Fine Chemicals, Uppsala, Sweden), which was totally excluded
from both columns.

The results were plotted on graphs with both the UV detection at 254nm
and cadmium content of the column eluate on the ordinate and elution volume
on the abscissa.   The UV absorption at 254nm enables all compounds containing an aromatic ring to be detected together with most aldehydes and
ketones whose absorption bands extend into this region.

Condensed aromatics can also be detected at this wavelength although
their principal absorption is at longer wavelengths.

3. RESULTS

The chromatographic fractionations for the soil solutions through a μ Bondagel E-500 HPLC column are shown in Figures 1 to 6. In each case it can be seen that the cadmium is associated with the lowest molecular weight UV absorption peak, irrespective of the source of cadmium contamination. No evidence of intermediate or high molecular weight cadmium-containing complexes was found even though various UV absorption peaks in these regions indicated the presence of larger organic molecules in all the solutions.

In view of the wide fractionation range of μ Bondagel E-500, it was decided to run the soil solution through a μ Porasil GPC 60Å column. This had a narrower range (100-10,000) and was expected to provide greater resolution of the lower molecular weight constituents. From the graphs in Figures 7 to 12 it can be seen that the cadmium was only associated with the largest UV absorbance peak although smaller-sized peaks occurred at higher molecular weight positions in all solutions, and a small UV peak appeared to elute after the main one in some of the soils. The peaks around Vt are due to solvent effects.

It is probable that the cadmium containing peak includes both organic and inorganic compounds of similar molecular weight, together with ionic cadmium (Cd^{2+}).

Earlier work with a cationic exchange resin indicated that a considerable amount of cationic cadmium existed in both mining-polluted and sludge-amended soils.[6]   Mahler et al (1980) estimated that free ionic cadmium accounted for 64-72% of the total cadmium in sludge-amended soils, using a computer model (GEOCHEM). [7]   They also found that organic complexes of cadmium usually comprised less than 10%, while SO_4, HCO_3 or CO_3 and Cl complexes, in decreasing order respectively, accounted for the inorganic forms of cadmium.

Opinions differ on the organic complexation of cadmium in soil solutions.   Holtzclaw et al (1978) found this element to be associated with the fulvic acid fraction of soil humus. [8]   However, Baham et al (1978), using gel permeation chromatography (Sephadex G10), observed that cadmium showed little tendency to associate with fulvic acid components.[9] Butterworth and Alloway (1981) found evidence of some higher molecular weight complexes using Sephadex gels.[10]   However, the different findings reported here (Figures 1-12) are considered to be more reliable owing to the superior resolution and precision of High Performance Liquid Chromatography.

From the graphs in Figures 1 to 12, it appears that cadmium may be associated with some relatively small organic molecules, possibly organic acids. It should be possible to separate these by reverse-phase chromatography and analyse the fractions to establish whether any of these compounds contain cadmium.

In addition to the characterization/identification of the cadmium species in the soil solution, many other aspects need to be investigated. These include changes in the speciation of the metal with time from application of sewage sludge and any variations between different sludges and soils. Other elements, such as lead are also being studied and preliminary results using gel permeation chromatography suggested that a much wider range of molecular weight complexes occur with lead. Even after identifying the metal species in soil solutions it will still be necessary to assess and compare the ease with which they are taken up by crop plants.

4. ACKNOWLEDGEMENTS

The authors gratefully acknowledge the receipt of a grant from the Department of the Environment (DGR/480/700 and PECD4/8/05).

5. REFERENCES

1. Peterson, P.J. and Alloway, B.J. (1979). Cadmium in soils and vegetation. Chapter 2 in 'The Chemistry, Biochemistry and Biology of Cadmium.' Ed. Webb, M.J., Elsevier, Amsterdam. 45-92.

2. Sposito, G. and Mattigod, S.V. (1977). On the chemical foundation of the sodium adsorption ratio. Soil Sci. Soc. Am. J., 41, 323-329.

3. Sposito, G. and Mattigod, S.V. (1980). GEOCHEM: A computer program for the calculation of chemical equilibria in soil solutions and other natural water systems. The Kearney Foundation of Soil Science, University of California.

4. Mattigod, S.V. and Sposito, G. (1979). Chemical modelling of trace metal equilibria in contaminated soil solutions using the computer program GEOCHEM. In Jenne, E.A. (Ed), Chemical Modeling in Aqueous Systems. ACS Symposium Series No. 93. American Chemical Society, Washington D.C. pp 837-856.

5. Alloway, B.J., Gregson, M., Gregson, S.K., Tanner, R. and Tills, A.R. (1979). Heavy metals in soils contaminated from several sources including sewage sludge. Proc. Int. Conf. Management and Control of Heavy Metals in the Environment. London 545-548.

6. Butterworth, F.E. and Alloway, B.J. (1981). Investigations into the speciation of cadmium in polluted soils using liquid chromatography. Proc. Int. Conf. Heavy Metals in the Environment. Amsterdam. 713-716.

7. Mahler, R.J., Binham, F.T., Sposito, G. and Page, A.K. (1980). Cadmium-enriched sewage sludge application of acid and calcareous soils: relation between treatment, cadmium in saturation extracts

and cadmium uptake. J. Environ. Qual. $\underline{9}$ (3) 359-364.

8. Holtzclaw, K.M., Keech, D.A., Page, A.L., Sposito, G., Ganje, T.J. and Ball, N.B. (1978).   Trace metal distributions among the humic acid, the fulvic acid, and precipitable fractions extracted with NaOH from sewage sludges. J. Environ. Qual. $\underline{7}$, 124-127.

9. Baham, J., Ball, N.B. and Sposito, G. (1978). Gel filtration studies of trace metal-fulvic acid solutions extracted from sewage sludges. J. Environ. Qual. $\underline{7}$ (2) 181-188.

Details of the Soils Used in the Soil Solution Studies

Sample	Site	Textural class*	%loss on ignition	Soil pH (water)	Soil solution pH	Soil total cadmium (μg g^{-1})
A	Disused sewage farm (City A)	SCL	26	5.9	4.74	16.1
B	Heavily sludged soil	SZL	25.6	5.8	6.75	88.3
C	Active sewage farm (City B)	CL	25.8	6.8	7.90	69.7
D	Rough pasture on the site of old Pb-Zn mine	SL	8.2	8.0	7.25	437.0
E	Control to C (no sewage but pig manure applied)	CL	7.5	6.6	7.36	0.1
F	Control to D (normal agricultural soil)	ZL	7.33	5.86	6.32	0.44

* Textural classes: SCL, sandy clay loam; SZL, sandy silt loam; CL, clay loam; ZL, silt loam.

Separation of soil solution using a u Bondagel E-500 column with 10ul injection. UV absorbance at 254nm 0.05a.u.f.s.(- - - -) and cadmium concentration (————) are plotted against elution volume.

SOIL A Disused Sewage
            Farm

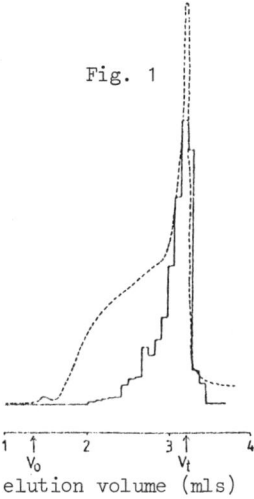

Fig. 1

elution volume (mls)

SOIL B Heavily
            Sludged

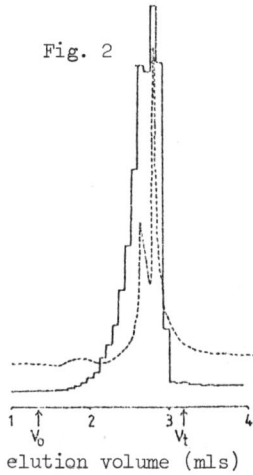

Fig. 2

elution volume (mls)

SOIL C Sewage Farm

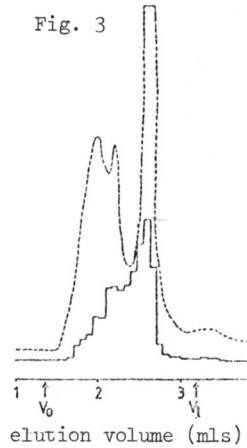

Fig. 3

elution volume (mls)

SOIL D Polluted
            Pb-Zn Mining

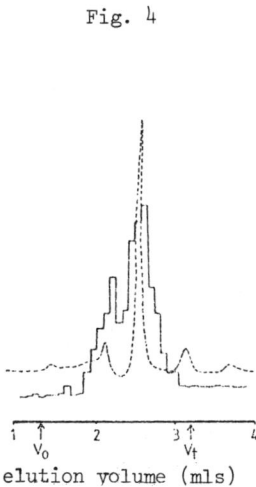

Fig. 4

elution volume (mls)

SOIL E Control to C

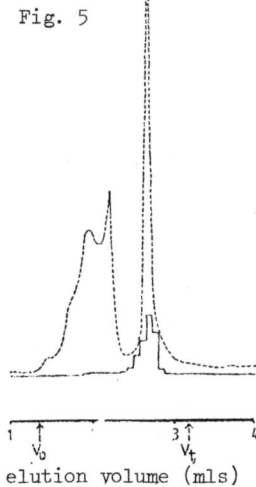

Fig. 5

elution volume (mls)

SOIL F Control to D

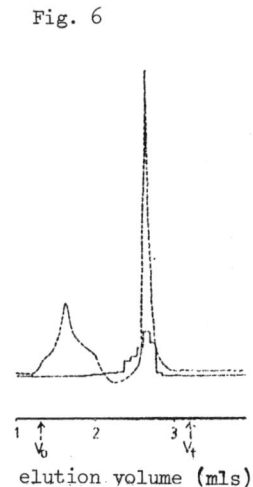

Fig. 6

elution volume (mls)

Separation of soil solution using a u Porasil GPC 60Å column with 20ul
injection. UV absorbance at 254nm 0.05a.u.f.s. (- - - -) and cadmium
concentration (———) are plotted against elution volume.

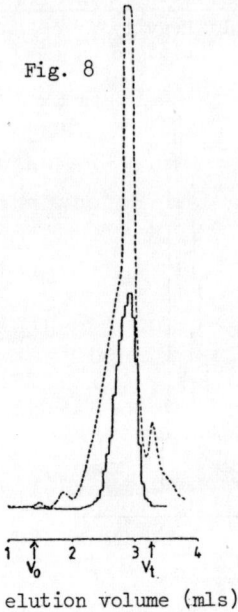

SOIL A Disused Sewage
                Farm

Fig. 7

elution volume (mls)

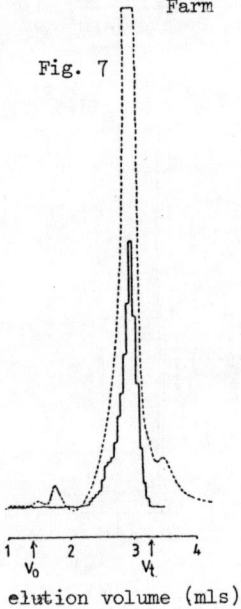

SOIL B Heavily Sludged

Fig. 8

elution volume (mls)

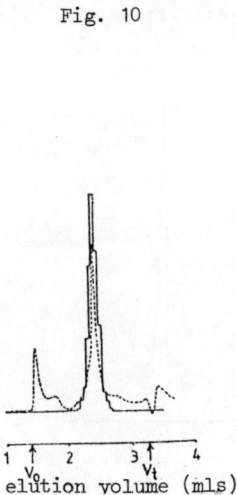

SOIL C Sewage
                Farm

Fig. 9

elution volume (mls)

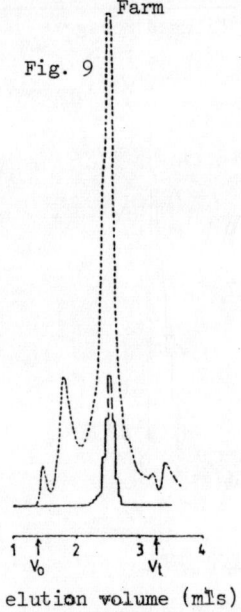

SOIL D Polluted
            Pb-Zn Mining

Fig. 10

elution volume (mls)

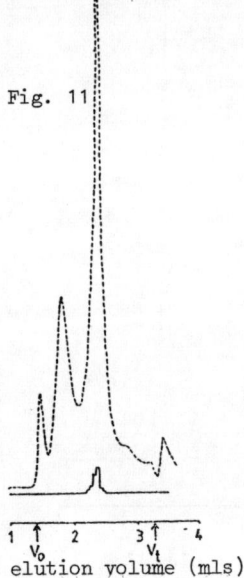

SOIL E Control to C

Fig. 11

elution volume (mls)

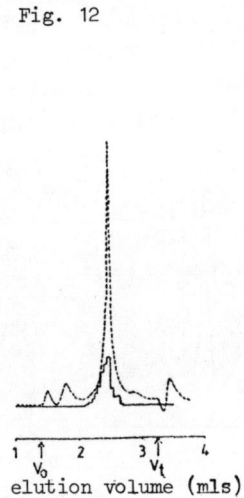

SOIL F Control to D

Fig. 12

elution volume (mls)

INFLUENCE OF DIFFERENT TYPES OF NATURAL ORGANIC

MATTER ON THE SOLUBILITY OF HEAVY METALS IN SOILS

U. HERMS and [+]G. BRÜMMER

Bodentechnologisches Institut Bremen and [+]Institut für
Pflanzenernährung und Bodenkunde Kiel

SUMMARY:

In model experiments, the pH-dependent solubility of Cd,
Zn, Cu and Pb was investigated in different soils. At pH
7 - 8 the organic matter of the soil often increases the
solubility of heavy metals, while at pH 4 - 6 there is an
immobilizing effect in relation to the influence of mineralic
substances. The addition of fresh and of slightly decomposed
organic material (hay) leads to an increasing solubility of
$Cu \gg Cd \gtrsim Zn > Pb$ at pH 4 - 8. On the other side, the addition
of strong decomposed organic matter (peat) brings about only
a slight mobilization at pH 8, but at pH 4 - 7 it leads to a
marked immobilization of $Cu > Cd > Zn > Pb$.

The results indicate, that after mixing of decomposable
organic matter (e.g. sewage sludge, manure) with soils an
increase in heavy metal solubility due to formation of
soluble, low molecular organic complexes can occur. After the
mobilization phase an immobilization of heavy metals takes
place. Further transformation of the added organic matter
leads to the formation of insoluble organic or organo-
mineralic compounds with a marked capacity to fix heavy
metals especially at $pH \leq 6$.

1. INTRODUCTION

Organic matter can influence the solubility of heavy

metals in soils in different ways. It can increase the solu-

bility by forming soluble organic complexes (1), but on the

other side the ability of organic matter to immobilize heavy

metals has also been reported (2). The mobilizing or immobi-

lizing effect of organic matter is determined 1. by soil

reaction, which influences the solubility of organic sub-

stances, 2. by the molecular weight of the formed complexes

(3) and 3. by the degradation state of organic compounds (1).

Most of the reported experiments on this subject were made in

simple model systems or with well defined organic substances,
but only few informations are available about the influence
of natural organic matter on the solubility of heavy metals
and the competitive effect of mineralic substances in soils.
This was the reason for us to carry out model experiments on
this subject with soil samples of different composition.
Some of the results are presented in this paper.

2. MATERIALS and METHODS

The soil samples were taken from A_p-horizons of different,
not-contaminated soils and from a subsoil horizon (Tab. 1).
All samples got additions of 100 mg/kg Cu, Zn and Pb and
15 mg/kg Cd. In a pre-incubation period of 8 weeks the
metals were equilibrated with the soil samples.

Tab. 1: General composition of the soil samples

No.	soil type	horizon	pH CaCl$_2$	C_{org} %	clay %	CaCO$_3$ %	Fe$_d$ %	CEC meq/100 g
1	podzol	A_p	5,0	2,2	2	–	0,13	18,7
2	brown earth	A_p	5,5	1,8	10	–	0,51	23,5
3	calcareous marsh soil	A_p	7,2	1,3	20	0,8	0,31	29,5
4	lessive	SB$_{vt}$	6,8	0,2	24	–	0,68	30,5

As model substances for natural organic compounds
additions of 1. 5% hay, 2. 5% peat or 3. and 4. 5% microbially activated hay and peat, respectively, were taken for
experiments with subsoil samples (soil 4, Tab. 1), which
contain a very low content of organic matter.

Samples of the soil material were suspended with water and
adjusted to different pH-values between 3 to 8. After an
equilibration period of two weeks at constant pH values and
sterile conditions, the concentrations of Cu, Zn, Cd and Pb
in the equilibrium solutions were determined by anodic
stripping voltammetry (Cd, Pb) or atomic absorption (Cu, Zn).

3. RESULTS and DISCUSSION

The pH-dependent solubility of Cd in A_p-horizons of
different soils with varying content of organic matter is

shown in Fig. 1. As expected, the solution concentration of
Cd decreases with increasing pH in the range of pH 3 to 7. At
pH values above 7, a further decrease takes place in soil 3
with 1,3% organic matter. In soil 2 with 1,8% organic matter
the Cd concentration remains constant at pH 7 to 8, whereas
a steep increase occurs in this pH range in soil 1 with 2,2%
organic matter. The mobilizing effect of organic matter at
neutral to alkaline reaction is due to increasing solubility
of chelating organic substances and also of organic metal
complexes. On the other side, at pH 3 to 6 the solubility of
Cd is lowest in soil 1, although this soil has the lowest
CEC of the soils taken for these experiments (Tab. 1). This
effect is due to the formation of insoluble organic or
organo-mineralic complexes with Cd at acid soil reaction.

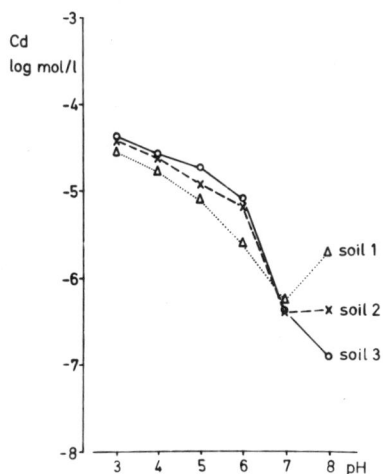

Fig. 1: Cd-content of soil
solutions after an equi-
libration period of two weeks
at different constant
pH-values

In a similar way, but to a different degree, a mobilization of other heavy metals takes place at neutral or
slightly alkaline conditions and an immobilization at acid
conditions. In general, the solubility of heavy metals is
influenced by the natural organic matter of soils in the
order $Cu > Pb > Cd \geq Zn$ (4), which is identical with the order
of decreasing stability of organic complexes (5).

In the almost humus free soil sample 4, the solubility of
Cd decreases strongly with increasing pH (Fig. 2). The

addition of 5% fresh organic matter (hay) to this soil leads
to a mobilization of Cd in the range from pH 4 to 8 (Fig. 2).
Somewhat less, but in relation to the control a mobilizing
effect also has the addition of 5% microbially activated
hay. A minimum of the Cd concentration in the equilibrium
solution at pH 7 is caused by addition of 5% peat to the
soil (Fig. 2). In relation to the control, peat slightly
immobilizes Cd in the range of pH 5 to 7, but mobilizes Cd at
pH above 7. The immobilizing effect of peat on Cd in inten-
sified by microbial activation of the peat. An addition of
5% activated peat brings about a strong immobilization of Cd
at pH 3 to 7,5 (Fig. 2).

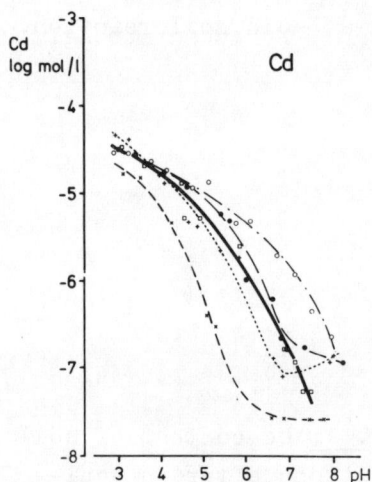

□——————□	without addition
o——— -o	+ 5% hay
●— —●	+ 5% microbially activated hay
+·······+	+ 5% peat
x— ——x	+ 5% microbially activated peat

Fig. 2: Influence of
different types of organic
matter on the pH-dependent
solubility of Cd in soil

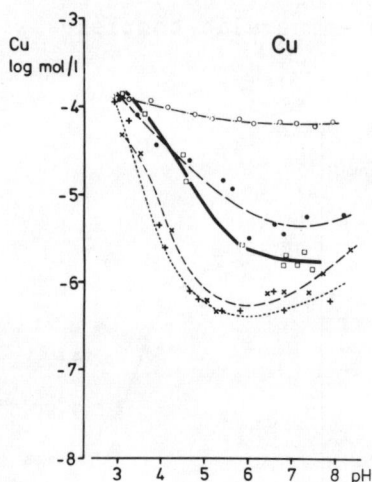

Fig. 3: Influence of different
types of organic matter on the
pH-dependent solubility of Cu
in soil

The solubility of Zn is influenced by organic compounds
in a similar way and to a similar extent as the solubility
of Cd.

Cu forms stronger complexes with organic substances than Cd or Zn (5). The effects of the different types of organic matter on the solubility of Cu in the soil are, therefore, much more pronounced than for Cd or Zn (Fig. 3). There is not only a strong mobilizing effect of fresh organic matter (hay) on Cu in the range of pH 4 to 8, but also a distinct minimum in the Cu solubility at pH 6 in samples containing peat and a remarkable depression of the Cu concentration at pH values below 6 in relation to the control (Fig. 3). This may be one reason for the fact, that Cu deficiency is often observed on acid humic soils.

Although Pb also forms rather stable organic complexes (5), its solubility is hardly affected by additions of the various organic compounds to the soil. Pb is so strongly fixed by mineralic soil components, that neither a further immobilization nor a marked mobilization of it can be caused by the addition of organic matter.

4. CONCLUSIONS

The results of these and other model experiments show, that mixing of fresh organic matter like hay, but also sewage sludge, manure or harvest residues with contaminated soils initially causes a mobilization of heavy metals in the order $Cu \gg Cd \geq Zn > Pb$ at pH 4 to 8. With further transformations of the added organic matter the mobilizing influence changes into an immobilizing effect, especially at acid soil reaction.

5. LITERATURE

(1) BLOOMFIELD, C., W.I. KELSO and G. PRUDEN, 1976: Reactions between metals and humified organic matter.
    Soil Sci. 27, 16 - 31
(2) STRICKLAND, R.C., W.R. CHANEY and R.J. LAMOUREAUX, 1979: Organic matter influences phytotoxicity of cadmium to soybeans.
    Plant and Soil 52, 393 - 402

(3) STEVENSON, F.J. and M.S. ARDAKANI, 1972: Organic matter
        reactions involving micronutrients in soils.
        In: Micronutrients in Agriculture, Soil Sci. Soc.
        Am., Inc., Madison, Wisc.
(4) HERMS, U. and G. BRÜMMER, 1980: Einfluß der Boden-
        reaktion auf Löslichkeit und tolerierbare Gesamt-
        gehalte an Nickel, Kupfer, Zink, Cadmium und Blei
        in Böden und kompostierten Siedlungsabfällen.
        Landwirtsch. Forschung 33, 408 - 423
(5) STEVENSON, F.J., 1977: Nature of divalent transition
        metal complexes of humic acids as revealed by a
        modified potentiometric titration method.
        Soil Sci. 123, 10 - 17

POSSIBILITIES OF REDUCING PLANT AVAILABILITY OF HEAVY METALS
IN A CONTAMINATED SOIL

L. KIEKENS & A. COTTENIE

Laboratory of Analytical and Agrochemistry
State University Ghent - Belgium

Summary

   The relative importance of different mechanisms such as precipitation, complexation and adsorption with regard to immobilization of heavy metals in a contaminated soil were investigated. Addition of a selective cation exchanger, a heavy clay soil or lime to a sandy soil reduced heavy metal availability in the soil as determined with various extractants. Heavy metal concentrations in corn plants were also decreased by increasing the cation exchange capacity or pH of the soil.
   Addition of peat to the soil only reduced availability of those metals which form stable organomineral complexes with soil organic constituents.
   Significant yield increases were obtained for all treatments, the sequence for decreasing effect being heavy clay soil > lime ≅ selective cation exchange resin > peat.
   Linear regression analysis across all treatments revealed significant correlations between heavy metal concentration in the plant and the sum of watersoluble, exchangeable and organically bound fractions of heavy metals in the soil.

1. INTRODUCTION

In recent years numerous investigations of the effects of elevated levels of heavy metals in soils have indicated that significant increases in metal concentrations of plant tissue as well as toxicities have resulted. The different metals were found to vary significantly in their phytotoxic effects. In addition, relative metal uptake and phytotoxicity were shown to be dependent on plant genotype as well as soil or substrate properties [HAGHIRI (18), COTTENIE & KIEKENS (7), PAGE (35), DOWDY & LARSON (14), BINGHAM et al. (2,3), GIORDANO & MAYS (17), DOYLE et al. (15), MITCHELL et al. (32), BINGHAM (4), DAVIS & CARLTON-SMITH (11), DAVIS & COKER (12), KIEKENS et al. (28), SMILDE (38)].

Numerous reports and papers have dealt with the effect of soil pH on availability and accumulation of heavy metals in plants. In general, increasing pH reduces plant uptake of heavy metals [HODGSON (21), LAGERWERFF (31), JOHN et al. (25), KING & MORRIS (30), COTTENIE & KIEKENS (7), GIORDANO et al. (16), BOLTON (6), CUNNINGHAM et al. (9), JONES et al. (26), WILLIAMS (43), MORTVEDT & GIORDANO (34), PAGE et al. (36), HYDE et al. (22), BINGHAM et al. (5), DAVIS & COKER (12), DIJKSHOORN et al. (13), ALLINSON & DZIALO (1)].

Another factor which influences the availability of potentially toxic metals added to the soil, is the cation exchange capacity. HAGHIRI (19) and MILLER et al. (33) showed that the Cd concentration in oat shoots and soybeans was decreased by increasing the CEC of the soil. Similar observations were made by KIEKENS et al. (28) with regard to Zn concentrations in corn shoots. In several papers [VAN ASSCHE & DE MEY (40), JANSEN & VAN ASSCHE (23,24), VAN ASSCHE & JANSSEN (41), HELLER (20)] the beneficial effects of adding selectively working cation exchangers to a contaminated soil on plant uptake of heavy metals have been demonstrated.

Organic matter also seems to have an ameliorating effect on the toxic potential of the metals present. CUNNINGHAM et al. (10) have found that toxic elements tended to be more available in the inorganic form than when present in sludge. SMEULDERS (37) showed that addition of synthetic complexing agents such as tetraethylenepentamine to the soil resulted in a decrease of plant availability of copper. It has generally been assumed that for an element to be taken up by a plant or to exert an effect on plant growth, it must be present in solution [STERRITT & LESTER (39)]. Thus, in cases of soil enrichment with heavy metals an immobilization of the soluble forms as irreversibly

as possible, seems to be one of the most indicated ways for reducing plant availability of heavy metals. For this immobilization one or more of the following processes may play a role :
- coagulation of suspended solids, precipitation or coprecipitation
- complexation and uptake in biological systems
- adsorption and lattice penetration.

The purpose of this paper was to study the relative importance of precipitation, adsorption and complexation with regard to immobilization of heavy metals in soils and their availability and uptake by plants.

2. MATERIALS AND METHODS

Equilibration and pot experiments were conducted in a greenhouse, using an acid sandy soil sampled near a metal-processing industrial plant. This soil had a pH of 4.1, a cation exchange capacity of 6.0 meq/100 g, an organic matter content of 2.7 % and respectively 31.5, 108, 0.59 and 650 ppm of aqua regia extractable Zn, Cu, Cd and Pb. These figures indicate that, in comparison to the normal levels found in Belgian soils [KIEKENS et al. (29)[the soil under study is enriched with copper and lead.
In this study, the following five treatments were used :
1. untreated sandy soil (blanc)
2. a mixture of sandy soil and cation exchange resin (ratio 50/1)
3. a mixture of sandy soil and peat (ratio 20/1)
4. a mixture of sandy soil and lime (ratio 250/1)
5. a mixture of sandy soil and heavy clay soil (ratio 5/1).

Each of the quantities of cation exchange resin, peat and clay soil added to the sandy soil increased the CEC of the soil with 4-5 meq/100 g. The cation exchanger used was Lewatit OC 1029 (Bayer, AG Leverkussen) a water-insoluble macroporous polystyrol resin loaded with Ca-ions. The heavy clay soil used in treatment 5 had a pH of 7.8 and a CEC of 25 meq/100 g. Previous experiments indicated that quite large quantities of heavy metals may be retained by this soil in unavailable form [KIEKENS (27)[.

For each treatment, six pots were filled with 1 kg of the soil mixture and brought to field capacity by adding distilled water. The following supplemental nutrients were mixed with the soil in each pot : 0.5 g NH_4NO_3, 0.25 g $Ca(H_2PO_4)_2$, 0.25 g K_2SO_4 and 0.25 g $MgSO_4$. For each treatment two pots were kept for the equilibration experiment and four pots were used in the pot trial.

Four maize plants were grown per pot for 6 weeks, during which the soils were kept at field capacity by adding daily deionized water. Supplemental lighting was provided to maintain a day length of 16 hours.

After harvest the forage in the four replicates of each treatment was dried, weighed and analyzed for Zn, Cu, Cd and Pb according to the methods described by COTTENIE et al. (8).
The soils from the equilibration experiment were air-dried and passed through a 2 mm sieve. Heavy metals in the soils were extracted with the following solutions :
- aqua regia, for the determination of total Zn, Cu, Cd and Pb
  (soil/solution ratio = 1/10, extraction time : overnight)
- water, for the determination of soluble Zn, Cu, Cd and Pb
  (soil/solution ratio = 1/5, 30 minutes)
- 1 N NH_4OAc pH = 7, for exchangeable heavy metals
  (soil/solution ratio = 1/10, 1 h)
- 0.2 N NaOH for the determination of the fraction of heavy metals bound to organic soil constituents (soil/solution ratio = 1/5, 30 min.).
In all soil extracts heavy metals were determined by AAS.

3. RESULTS AND DISCUSSION

3.1. Soil analysis

Total, watersoluble, exchangeable and organically bound heavy metals extracted from the differently treated soils are summarized in table I.

Addition of cation exchanger, heavy clay soil or lime to the sandy soil generally resulted in a marked decrease of watersoluble, exchangeable and 0.2 N NaOH extractable fractions of Zn, Cu, Cd and Pb. Addition of peat particularly reduced the exchangeable copper fraction. This may be due to the higher affinity of soil organic constituents (fulvic and humic acids) to form with copper stable organomineral complexes [VERLOO (42)[. This is confirmed by the amounts of heavy metals extracted with 0.2 N NaOH, which is a measure for the organically bound fraction. The highest amounts of heavy metals extractable with 0.2 N NaOH were found in the peat treatment, indicating that in this case heavy metals are mainly involved in complexation reactions. For the other treatments as well precipitation as adsorption reactions play a more important role in the immobilization of heavy metals in the soil. The most significant effect was obtained in the heavy clay soil treatment, due to a combined action of irreversible adsorption (hysteresis) and pH increase. Indeed, the pH value of the soil treated with heavy clay soil showed an increase of about 2 units over the pH of the other treatments, except lime.

A pH increase seems to be particularly effective for the immobilization of lead. For this element the most significant effect was obtained in the lime and heavy clay soil treatments. Now, the question arises whether the results of soil analysis are reflected in plant response.

Table I. Concentrations of Zn, Cu, Cd and Pb in the treated soils after equilibration as determined with four different extractants

Soil treatment	extracting solution	concentration (mg/kg soil, ppm)			
		Zn	Cu	Cd	Pb
I[+]	aqua regia	31.5	108	0.59	650
II		32.4	117	0.57	668
III		33.4	125	0.50	686
IV		31.8	118	0.68	641
V		42.9	117	0.68	682
I	H_2O	0.60	0.15	0.13	1.3
II		0.40	0.15	0.04	0.7
III		0.30	0.30	0.07	1.5
IV		0.15	0.01	0.04	1.6
V		0.03	0.03	0.02	0.2
I	1 N NH_4OAc pH=7	2.03	7.7	0.08	92.5
II		0.13	1.4	0.03	53.8
III		2.30	0.7	0.09	81.7
IV		0.40	2.7	0.08	50.5
V		0.29	1.7	0.08	10.8
I	0.2 N NaOH	3.4	92	0.21	219
II		1.2	43	0.13	173
III		3.3	75	0.16	198
IV		1.1	55	0.11	91
V		0.9	25	0.05	21

[+]Soil treatments : I = blanc ; II = soil + cation exchanger ; III = soil + peat ; IV = soil + lime ; V = soil + heavy clay soil

3.2. Plant Response

Table II shows the dry matter yields of corn plants and the total plant concentrations of heavy metals.

Table II. Dry matter yield and concentrations of heavy metals in corn plants as affected by the different soil treatments

Soil treatment	Yield (g dry matter)	concentration in plant (mg/kg DM)			
		Zn	Cu	Cd	Pb
I Untreated (blanc)	1.10	152.7	59.5	1.10	88.2
II + Cation exchange resin	3.79	34.9	11.0	0.45	27.7
III + peat	2.15	114.5	27.9	0.68	51.0
IV + lime	3.76	25.4	15.6	0.45	4.7
V + heavy clay soil	6.57	20.7	9.4	0.10	3.6

All soil treatments gave significant yield increases over control, ranging from 200 % for the peat treatment to 600 % for the heavy clay soil treatment (fig. 1). Similar increases of about 250 % were obtained for both the cation exchange resin and the lime treatment. Thus it seems that quite equal effects were obtained by increasing the pH and by increasing the CEC of the sandy soil. A combination of both mechanisms in the heavy clay soil treatment gave the most significant effect.

For the different treatments a significant decrease of the metal concentrations in plants was observed (fig. 1). The effect of the treatments on concentration decrease of Zn, Cu and Cd, showed the following sequence : clay soil > lime $\cong$ cation exchange resin > peat. The concentration of Pb in corn plants was significantly different for the lime and cation exchange resin treatment, the most significant decrease being obtained with lime.
This may be explained by the fact that of the four elements under study Pb and Cu will begin to precipitate at lower pH values than Cd and Zn. For Cu however no significant difference was observed between lime and cation exchange resin. This is due to the higher affinity of Cu to combine with the chelating groups of the cation exchange resin. The importance of complexation with regard to Cu is also shown by the fact that of the four elements studied the peat treatment was most effective for Cu.

Table III shows the correlation obtained from linear regression analysis across all treatments of plant concentrations of heavy metals with heavy

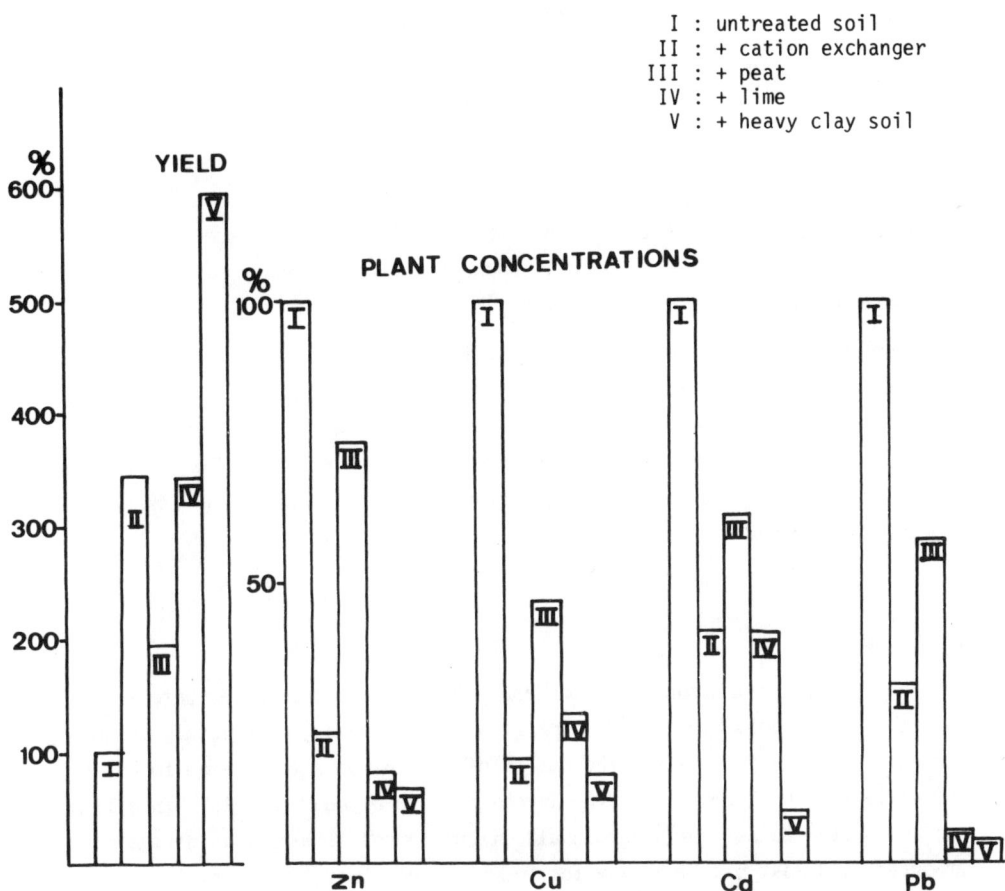

Fig. 1. Relative yields and concentrations of heavy metals in corn plants for the different treatments

metal fractions in the soil.

Table III. Correlation coefficients (r) from linear regression analysis across
all soil treatments of plant concentrations of heavy metals with
heavy metal content of the soil as determined with different
extractants

Soil content	Plant concentration			
	Zn	Cu	Cd	Pb
watersoluble (H_2O)	0.777	0.372	0.976[***]	0.839[*]
exchangeable (1 N NH_4OAc pH = 7)	0.936[**]	0.860[*]	0.186	0.875[*]
organically bound (0.2 N NaOH)	0.980[***]	0.904[*]	0.983[***]	0.862[*]
Σ (H_2O + 1 N NH_4OAc pH 7 + 0.2 N NaOH)	0.977[***]	0.929[**]	0.980[***]	0.876[*]

*, **, *** indicate 0.05, 0.01, 0.001 levels of significance respectively.

Significant correlations of plant concentrations with the sum of water-soluble, exchangeable and organically bound fractions of heavy metals in the soil were obtained for all elements.

Plant Cu concentration was most significantly correlated with 0.2 N NaOH extractable soil Cu. Plant Pb concentration was significantly correlated with all extractable soil Pb fractions. The most significant correlations for plant Zn and Cd concentrations were obtained with the 0.2 N NaOH extractable fractions, and also significant correlations were found with respectively exchangeable Zn and watersoluble Cd.

4. CONCLUSIONS

The results of the equilibration and pot experiment presented here indicate that it is possible to reduce plant availability of heavy metals in an enriched soil. An increase of the number of adsorption sites seemed to be as effective as a pH increase for the immobilization of heavy metals. Since lime can be applied to all soils and in all circumstances, liming remains the most important and most practical way for reducing heavy metal availability in soils.

A second and also important mechanism is adsorption. The larger the adsorption-desorption hysteresis effect, the more important becomes this mechanism. This is only possible at sufficient number of adsorption sites and this effect will be dominant in heavy soils with high CEC values. In light textured soils with low CEC values, the effect of adsorption may be enhanced by increasing the CEC value of the soil ; which may be achieved by addition of clay or, to a lesser extent, by addition of organic matter.

A combination of both, pH increase (liming) and increase of CEC (addition of clay) seems to be the most effective way for immobilization of heavy metals in the soil and for reducing their plant uptake.

ACKNOWLEDGEMENTS

The author wishes to thank the "Instituut tot Aanmoediging van het Wetenschappelijk Onderzoek in Nijverheid en Landbouw" (I.W.O.N.L.) for financial assistance.

REFERENCES

1. Allinson, D.W. and Dzialo, C. 1981. The influence of lead, cadmium and nickel on the growth of ryegrass and oats. Plant and Soil 62 : 81-89.
2. Bingham, F.T., Page, A.L., Mahler, R.J. and Ganje, T.J. 1975. Growth and cadmium accumulation of plants grown on a soil treated with a cadmium-enriched sewage sludge. J. Environ. Qual. 4 : 207-211.
3. Bingham, F.T., Page, A.L., Mahler, R.J. and Ganje, T.J. 1976. Yield and cadmium accumulation of forage species in relation to cadmium content of sludge-amended soil. J. Environ. Qual. 5 : 57-60.
4. Bingham, F.T. 1979. Bioavailability of Cd to food crops in relation to heavy metal content of sludge-amended soil. Environ. Health Perspectives 28 : 39-73.
5. Bingham, F.T., Page, A.L., Mitchell, G.A. and Strong, J.E. 1979. Effects of liming an acid soil amended with sewage sludge enriched with Cd, Cu, Ni and Zn on yield and Cd content of wheat grain. J. Environ. Qual. 8 : 202-207.
6. Bolton, J. 1975. Liming effects on the toxicity to perennial rye-grass of a sewage sludge contaminated with zinc, nickel, copper and chromium. Environmental Pollution 9 : 295-304.

7. Cottenie, A. and Kiekens, L. 1974. Quantitative and qualitative plant response to extreme nutritional conditions, p. 543-556. *In Plant Analysis and fertilizer problems*. Proc. 7th IPI Colloquium Hannover, Fed. Rep. Germany.

8. Cottenie, A., Verloo, M., Kiekens, L., Velghe, G. and Camerlynck, R. 1982. Chemical analysis of plants and soils. Lab. Anal. Agrochemistry, State University Ghent, Belgium, p. 63.

9. Cunningham, J.O., Ryan, J.A. and Keeney, D.R. 1975. Phytotoxicity in and metal uptake from soil treated with metal-amended sewage sludge. J. Environ. Qual. 4 : 455-460.

10. Cunningham, J.O., Keeney, D.R. and Ryan, J.A. 1975. Yield and metal composition of corn and rye grown on sewage sludge amended soil. J. Environ. Qual. 4 : 448-454.

11. Davis, R.D. and Carlton-Smith, C. 1980. Crops as indicators of the significance of contamination of soil by heavy metals. Technical Report TR 140, Stevenage Laboratory. Water Research Centre, p. 44.

12. Davis, R.D. and Coker, E.G. 1980. Cadmium in agriculture, with special reference to the utilisation of sewage sludge on land. Technical Report TR 139, Stevenage Laboratory. Water Research Centre, p. 112.

13. Dijkshoorn, W., Lampe, J.E.M. and Van Broekhoven, L.W. 1981. Influence of soil pH on heavy metals in ryegrass from sludge-amended soil. Plant and Soil 61 : 277-284.

14. Dowdy, R.H. and Larson, W.E. 1975. The availability of sludge-borne metals to various vegetable crops. J. Environ. Qual. 4 : 278-281.

15. Doyle, P.J., Lester, J.N. and Perry, R. 1978. Survey of literature and experience on the disposal of sewage sludge on land. Final Report to Department of the Environment, London.

16. Giordano, P.M., Mortvedt, J.J. and Mays, D.A. 1975. Effect of municipal wastes on crop yields and uptake of heavy metals. J. Environ. Qual. 4 : 394-399.

17. Giordano, P.M. and Mays, D.A. 1977. Effect of land disposal applications of municipal wastes on crop yields and heavy metal uptake. USEPA Technol. Ser;., EPA-00/2-77-014.

18. Haghiri, F. 1973. Cadmium uptake by plants. J. Environ. Qual. 2 : 93-96.

19. Haghiri, F. 1974. Plant uptake of cadmium as influenced by cation exchange capacity, organic matter, zinc and soil temperature. J. Environ. Qual. 3 : 180-183.

20. Heller, H. 1979. Uber die Verwendung selektiv wirkender Kationenaustauscharze zur Festlegung phytotoxischer Schwermetalle in Kulturböden. Landwirtsch. Forsch. 32 : 138-149.

21. Hodgson, J.F. 1963. Chemistry of the micronutrient elements in soils. Advances in Agronomy 15 : 119-160.

22. Hyde, H.C., Page, A.L., Bingham, F.T. and Mahler, R.J. 1979. Effect of heavy metals in sludge on agricultural crops. Journal WPCF 51 : 2475-2486.

23. Jansen, G. and Van Assche, C. 1977. Results of cation exchangers against heavy metals in the field. Med. Fac. Landbouww. Rijksuniv. Gent, 42 : 1127-1134.

24. Jansen, G. and Van Assche, C. 1978. The phytiatric influence of treatments with cation exchangers in a soil, contaminated with heavy metals, on the growth and the mineral composition of Zea mays and Phaseolus vulgaris L. Plant and Soil, 49 : 229-249.

25. John, M.K., Van Laerhoven, C.J. and Chuah, H.H. 1972. Factors affecting plant uptake and phytotoxicity of cadmium added to soils. Environ. Sci. Technol. 6 : 1005-1009.

26. Jones, R.L., Hinsley, T.D., Ziegler, E.L. and Tyler, J.J. 1975. Cadmium and zinc contents of corn leaf and grain produced by sludge-amended soil. J. Environ. Qual. 4 : 509-514.

27. Kiekens, L. 1980. Adsorptieverschijnselen van zware metalen in gronden. Doctoral Thesis. State University Ghent, Belgium, p. 166.

28. Kiekens, L., Verloo, M. and Cottenie, A. 1980. Uptake of zinc by corn as influenced by clay and humic acid content of the substrate. In Proc. 5th Intern. Coll. Control Plant Nutrition, Castelfranco Veneto, Italy.

29. Kiekens, L., Verloo, M. and Cottenie, A. 1981. Behaviour and biological importance of trace elements in the soil. In Trace elements in agriculture and in the environment, Ed. A. Cottenie, pp. 3-24.

30. King, L.D. and Morris, H.D. 1972. Land disposal of liquid sewage sludge. II. The effect on soil pH, manganese, zinc and growth and chemical composition of rye (Secale cereale L.). J. Environ. Qual. 1 : 425-429.

31. Lagerwerff, J.V. 1971. Uptake of cadmium, lead and zinc by radish from soil and air. Soil Sci. 111 : 129-133.

32. Mitchell, G.A., Bingham, F.T. and Page, A.L. 1978. Yield and metal composition of lettuce and wheat grown on soils amended with sewage sludge enriched with cadmium, copper, nickel and zinc. J. Environ. Qual. 7 : 165-170.

33. Miller, J.E., Hassett, J.J. and Koeppe, D.E. 1976. Uptake of cadmium by soybeans as influenced by soil cation exchange capacity, pH, and available phosphorus. J. Environ. Qual. 5 : 157-160.

34. Mortvedt, J.J. and Giordano, P.M. 1975. Response of corn to zinc and chromium in municipal wastes applied to soil. J. Environ. Qual. 4 : 170-174.

35. Page, A.L. 1974. Fate and effects of trace elements in sewage sludge when applied to agricultural lands. A literature review study. Environ. Prot. Technol. Ser. EPA-670/2-74-005, p. 96.

36. Page, A.L., Chang, A.C. and Bingham, F.T. 1979. Trace metal phytotoxicity and absorption by crop plants grown on sewage sludge amended soils. Proc. Intern. Conf. on Management and control of heavy metals in the environment. Imperial College, London, pp. 525-528.

37. Smeulders, F. 1980. In situ immobilization of transition metal ions in soils with tetraethylenepentamine. Doctoral thesis. K.U.L. Belgium.

38. Smilde, K.W. 1981. Heavy-metal accumulation in crops grown on sewage sludge amended with metal salts. Plant and Soil, 62 : 3-14.

39. Sterritt, R.M. and Lester, J.N. 1980. The value of sewage sludge to agriculture and effects of the agricultural use of sludges contaminated with toxic elements : a review. Science Tot. Environ. 16 : 55-90.

40. Van Assche, C. and De Mey, W. 1975. Fytiatrie ten opzichte van de zware metalen met kationenuitwisselaars. Genootschap Plantproductie- en Ecosfeer, K.VIV, Gent.

41. Van Assche, C. and Jansen, G. 1978. Anwendung von selektiv wirkenden Kationenaustauschern auf mit Schwermetallen kontaminierten Böden. Landwirtsch. Forsch., 34 : 215-228.

42. Verloo, M.G. 1974. Komplexvorming van sporenelementen met organische bodemkomponenten. Doctoral thesis. State University Ghent, Belgium.

43. Williams, J.H. 1975. Use of sewage sludge on agricultural land and the effects of metals on crops. J. Water Pollut. Control, 74 : 635-644.

GEOCHEMICAL POLLUTION - SOME EFFECTS ON THE SELENIUM AND MOLYBDENUM CONTENTS OF CROPS

G. A. FLEMING

An Foras Taluntais
Johnstown Castle Research Centre
Wexford, Ireland

Summary

Soil pollution can result from natural processes as well as from human activity. Some rock types e.g. black shales accumulate relatively high levels of a number of heavy metals and soils formed from them become enriched to the point where they can produce toxic vegetation. In Ireland some soils are enriched with selenium and molybdenum and problems with grazing stock have been encountered. The capacity of a number of crops to accumulate selenium and molybdenum was studied in a pot trial with toxic soil. All crops grown contained high levels of both elements with *leguminosae* and *cruciferae* accumulating the greatest amounts. The uptake of selenium and molybdenum paralleled that of sulphur. With regard to the land application of materials such as sewage sludge it is necessary to be aware of naturally (geochemically) polluted areas in order to ensure that further additions of unwanted metals are not made.

1. INTRODUCTION

Pollution of the environment by heavy metals is normally associated with human activity of one kind or another. One aspect concerns the addition of unwanted materials to soils and in the case of heavy metals this may arise in a number of ways. The following may all contribute:-

- atmospheric fallout in industrial regions
- erosion of mine spoil heaps
- pesticides and fungicides
- dumped wastes
- pig slurries
- sewage sludges

Man however is not the sole polluter. Natural processes are capable of polluting our soils and consequently the food we eat. In the vicinity of ore bodies for instance soils and streams are frequently enriched with heavy metals. Lead pollution of soil resulting from the erosion of a galena ore body by stream water and subsequent deposition in a depression has been reported from Norway (1). This has resulted in barren soils unable to support vegetation other than some lead tolerant species.

An interesting case of molybdenum toxicity affecting humans has been recorded in India (2). The condition known as genu valgum, in which there can be severe distortion of the knee joints, occurs mainly in young adults and is associated with high intakes of both molybdenum and fluorine. The increase in molybdenum levels in the diet is held to have resulted from the higher water table and the more alkaline soil conditions which have resulted following the construction of a large dam. Here we have a soil pollution effect resulting from changing soil conditions but precipitated initially by man's activity.

Enrichment of soils with selenium and molybdenum has occurred in Ireland and constitutes true "geochemical pollution". Toxic soils are found mainly in counties Limerick, Tipperary and Meath. The Irish situation is by no means unique - seleniferous and molybdeniferous soils occur in many parts of the world. Large areas of seleniferous soils occur in the US while in the UK the "teart pastures of Somerset" while not high in selenium, are well known for their high content of molybdenum with resultant animal production problems.

2. FORMATION OF SELENIFEROUS AND MOLYBDENIFEROUS SOILS - BLACK SHALES

As the majority of seleniferous soils and some molybdeniferous ones are associated with the occurrence of black shales
it is necessary to deal briefly with the formation of this
rock facies. Apart from selenium and molybdenum black shales
may be enriched with a number of heavy metals including arsenic,
copper, lead, vanadium and uranium. Perhaps one of the best
known is the Kupferschiefer of continental Europe, which is of
Permian age and highly enriched with copper. The rocks
however may be of any geological age. In the US, black shales
of Cretaceous age contain   high levels of selenium and
molybdenum while in the UK black shales of Cretaceous age occur
in ·SE England and of Jurassic age in Yorkshire (3). The black
shales owe their colour to organic matter and they may form
under a variety of aquatic environments, including lagoons,
deep land locked sea basins and sea bottom basins. There is
one common factor however - the rate of deposition of organic
matter exceeds its oxidation or mineralization. Black shales
are therefore formed in an essentially anaerobic milieu and
result from the compression and diagenesis of organic muds.

They are characteristically rich in sulphides. With
regard to selenium and molybdenum their accumulation is
intimately associated with their affinity for both organic
matter and sulphur. Selenium under anaerobic conditions is
reduced to selenide, while sulphur is reduced to sulphide.
Because of the similarity in size of the selenide and sulphide
ions selenium will be found as a guest element in e.g. ferrous
sulphide (FeS). Molybdenum is frequently found as the mineral
molybdenite (MoS_2). Where black shales occur sufficiently
close to the surface to form the parent material of soils
their weathering releases selenium and molybdenum which will
then concentrate in depressions such as peaty swamps.   In
Ireland black shales of Namurian (mid-Carboniferous) age
occur in association with limestone and under the mildly acid
to alkaline conditions prevailing the mobility of both selenate
and molybdate ions is increased thus aiding their transport to
low lying areas. The net result is an organic soil containing
toxic levels of both selenium and molybdenum. Heavy-textured

mineral soils may also have enhanced levels of selenium and
more particularly molybdemum, but the most toxic soils are
invariably low-lying poorly drained and high in organic matter.

3. SELENIUM AND MOLYBDENUM IN PLANTS

Samples of field crops and herbage taken from one area of
toxic soils showed that levels of selenium were quite high.
Swedes' roots contained 5 μgg^{-1} Se with 30 μgg^{-1} in the leaves.
Sugar beet and carrots followed the same pattern i.e. more Se
in the leaves than in the roots while cabbage leaves contained
118 μgg^{-1} and roots had 37 μgg^{-1}. Herbage in some cases
contained over 100 μgg^{-1} Se.

In order to examine as many crops as possible, soil was
collected from one of the toxic fields and used in a pot trial.
The trial was conducted in the open in a bird free cage. The
soil had a pH of 6.8: organic matter content was 40%: total
Se 161 μgg^{-1} and total Mo 50 μgg^{-1} (4). Plants were grown in
Mitscherlich pots of 22.5 cm diameter. Five pots of each
species were sown and harvested material bulked for analysis.

The contents of selenium, molybdenum and sulphur found
in the various plants are shown in Table I. Levels of all
three elements were highest in the *leguminosae* and *cruciferae*
and where leaves and roots are compared, leaves almost
invariably contained the greater amounts. Sulphur levels are
important from the animal nutrition point of view because high
levels exacerbate molybdenum toxicity. In practice high levels
of molybdenum in an animal's diet can induce a low copper
status. This is thought to result from the chelation of
copper by thiomolybdates thus reducing its availability to
the animal. The importance of sulphur thus becomes apparent.
Young cattle are particularly at risk from molybdenum toxicity.

From the human nutrition point of view the levels of
selenium and molybdenum in the edible portions of crops is of
obvious importance. Cereal grains contained similar levels
of selenium to straw, but in the case of molybdenum, grains
contained approximately half as much. Lettuce had about twice
as much selenium as grasses and much less molybdenum. Turnips,
cabbage and onions are probably the most significant from the
human nutrition point of view but here it must be stressed that
the effect of cooking has not been taken into consideration. The

levels of both selenium and molybdenum in cress were quite high
but this must be seen in the context of the proportion of this
vegetable in a normal diet.

4. CONCLUSION

Soil pollution is normally associated with human activity
but situations do exist where soil forming processes combined
with the presence of certain rock types have conspired to
create areas of "natural" or "geochemical" pollution. While
such areas are certainly the exception rather than the norm,
their possible presence should be borne in mind when additions
of materials such as sewage sludge are contemplated. The case
history cited above deals only with two toxic heavy metals,
selenium and molybdenum but both are possible constituents of
sewage sludge. Toxicities of both elements are more likely
to affect animals rather than humans but because they can be
taken up in relatively large quantities by vegetables, their
addition to soils used for market gardening or private allot-
ments would not be encouraged.

5. REFERENCES

1) Lag, J., Hvatum, O.Ø und Bølviken, B (1969). An occurrence
    of naturally lead-poisoned soil at Kastad near Gjøvik,
    Norway. Norges Geologiske Undersøkelse No 266 141-159.

2) Agarwal, A.K. (1975). Crippling cost of India's big dam.
    New Scient. 65: 260-261.

3) Dunham, K.C. (1961). Black shale, oil and sulphide ore.
    Adv. Sci. 18: 284-299.

4) Fleming, G.A. (1962). Selenium in Irish soils and plants.
    Soil Sci. 94: 28-35.

Table I: Selenium, molybdenum and sulphur content of plants
          grown on toxic soil*

Species	Se ($\mu g g^{-1}$)	Mo ($\mu g g^{-1}$)	S %
GRAMINIAE			
Cocksfoot	28	39	.21
Perennial ryegrass	34	26	.23
Wheat - grain	39	16	.14
straw	40	42	.26
Oats - grain	39	24	.20
- straw	40	44	.26
Barley - grain	35	16	.14
- straw	42	35	.25
LEGUMINOSAE			
White clover	153	400	.49
Red clover	103	225	.45
Peas - leaves	79	31	-
- peas	9	-	-
COMPOSITAE			
Lettuce	56	7	.30
Artichoke - leaves	71	46	.17
- roots	19	5	.04
CRUCIFERAE			
Radish - leaves	145	55	-
- roots	35	16	.27
Cress - leaves	212	407	-
Cabbage - leaves	196	174	.46
Turnip - leaves	409	192	-
- roots	204	96	.19
Rape - leaves	203	276	.56
LILIACEAE			
Onion - leaves	235	240	.19
- bulbs	82	33	.21
UMBELLIFERAE			
Carrot - leaves	29	16	.31
- roots	-	-	.09
Parsnip - leaves	90	30	.29
- roots	22	3	.07

*Contents calculated on oven dry weight

EFFECTS OF HEAVY METALS ON SOIL MICROORGANISMS[*]

S. COPPOLA

Istituto di Microbiologia agraria e Stazione di Microbiologia industriale
Università di Napoli, 80055 Portici, Italia

Summary

Biotic and abiotic actions can affect heavy metals toxicity towards microorganisms in the soil environment.

Soil organic matter, clay minerals, chelating agents, H_2S resulting from microbial metabolism, different ions, pH and other factors may be responsible for the biologically effective concentration of metals. Moreover many microorganisms can tolerate heavy metals, producing cell surface structures or by genetically determined resistence, which controls cell uptake and permeability or the ability to transform the metal into an innocuous form.

A lot of incubation studies, in vitro researches and field experiments have shown that heavy metals can adversely affect mineralization processes, nitrification and symbiotic nitrogen fixation, mainly in acid soils.

Experiments with a volcanic soil and a sample of "terra rossa" have pointed out the influence of Cadmium upon ammonification, nitrification and non-symbiotic dinitrogen fixation, showing that the metal effect may also depend on the microbiological characteristics of each soil.

[*] This study was supported by the Consiglio Nazionale delle Ricerche, Project: Promozione della qualità dell'ambiente, Roma.

As man discovered chemical, physical and technological properties of heavy metals,greater and greater quantities of these elements were extracted from rocks and inserted into the biosphere, far exceeding the rates of natural cycling processes.On the other hand biological systems have not been able to evolve structures and mechanisms entirely agreable with the properties of all the elements (about 40) with a density greater than five. Thus some heavy metals are also reported to have comprehensively toxic effects on cells, mainly as a result of their ability to interact with biological polymers and making these molecules physiologically ineffective.

Within the utilization of sewage sludge and other organic wastes on agricultural lands heavy metals represent an ecotoxycological risk not only for food and feed produced by treated soils, but also for those adverse influences they can exert on microbial activities conditioning soil fertili ty. This last feature appears of particular interest, since agricultural lands are already tolerating extensive use of heavy metals in fungicides, disinfectants and, as contaminants, in fertilizers.

Soil microorganisms are indeed affected by heavy metals as the result of a multiplicity of interactions that can occur between microbial cells, ions and other environmental constituents (11).It is already known that metal availability in the soil is limited by binding with humic substances as well as with clay particles and it finally depends on the CEC (12,20,32). The clay minerals montmorillonite and kaolinite protected microorganisms, including bacteria, actinomycetes and filamentous fungi from inhibitory effects of Cd and the protective ability of clays was correlated with their CEC (3). Furthermore, prevention of toxic effects can be attributed to the presence of chelating agents (24,41) or to precipitation with H_2S resulting either from bacterial reduction of sulfates (18) or from mineralization of sulfur organic compounds (33). Metal toxicity is then pH dependent,general ly potentiated at pH 8 or 9 (2),and widely affected by other ions with simple additive effect (40), antagonism (1) or for the presence of anions (29) in the microbiological medium. Finally the biologically effective concentration may be reduced by formation of complexes between metal ions and surface structures, as like as capsules or outer layers of microbial cells (16).

But in addition to the above mentioned actions, authentic microbial resistance to heavy metals may be also genetically determined, depending on the ability of some microorganisms to biochemically transform toxic metals and to control cell uptake and permeability. Biogeochemical cycles of Hg (39) and of Sn (26) are far supported by microbial activity. Saxena and Howard (30) reviewed environmental transformations of alkylated and inorganic forms of various metals and pointed out experimental evidences concerning the microbial ability to support metal changes of valency and methylation reactions. Jeffries (15) recently reviewed the microbiology of Mercury, reporting both the ability of microorganisms to mobilize, concentrate, methylate, demethylate, volatilize Hg, and the first instance of application of microbial process to detoxification of industrial waste streams. At last, true resistance to some heavy metals is resulted in bacteria to be controlled by genes on extrachromosomal resistance (R) factors (plasmids) (31); the plasmid-mediated resistance depending on decreased uptake or on the ability of the organism to transform the metal into an innocuous form (6,7,17).

Such a variety of biotic and abiotic actions that can affect metal toxicity towards microorganisms in the natural environment does not allow to generalize about the effects each total concentration of toxic metal may produce on soil microorganisms. Microorganisms are often neglected in evaluation of the total detrimental effects of anthropogenic contaminants on the biosphere (4). At any rate some information remains interesting. The adverse effect of Cd added to soil was reflected in decreased microbial populations and depressed respiration rates (9). 10 ppm of Cd (as $CdCl_2$) produced a 41% reduction of O_2 consumption and a 36% decrease of CO_2 evolution from coniferous forest soil/litter microcosms (5). Decomposition studies of spruce litter from sites around two metal-processing industries in Sweden,which were emitting Cu, Zn, Cd, Ni and Pb,showed lower biological activity than comparable litter sample from a nonpolluted area (28). Soil phosphatase and urease appeared much reduced too (34). Decomposition rate, P mineralization and phosphatase activity decreased with increasing degree of Cu and Zn pollution in forest soils, appearing edaphically poor sites more sensitive to heavy metals pollution (36).In acid forest soils, Cu and Zn concentrations of three times the background level, were sufficient to

bring about a measurable disturbance of the nitrogen mineralization rate
(35). Cd is reported to be differently tolerated by soil fungi, to depress
micelial growth and affect fungus-plant,bacteria-plant and bacteria-fungus
interactions (4). Moreover Cd, and at a lower extent Zn and Pb,were found
to exhibit inhibition of growth of nematode-trapping fungi; inhibition
directly correlated with a decreased capacity to form traps (27). As far as
specific soil microbial activities, in vitro experiments have shown that
12.25 ppm of Cd completely inhibit nitrogen fixation, nitrification and the
hydrolysis of starch,whereas proteolysis,ammonification and denitrification
were strongly depressed.5.1ppm of Cu inhibited nitrification and adversely
affected the other activities (21). Nitrification rates resulted increased
in clayey mull soils incubated with 9÷18 µM $CdCl_2$ or 9÷22 µM Cd-acetate x g
of soil (37).At a similar extent,121 µM xg of soil of Pb-acetate significan
tly increased nitrate accumulation.Ammonium utilization and nitrification
were unaffected in fine sandy loam soil amended with 200 ppm of Cadmium.
Addition of either 500 or 1,000 ppm temporarily depressed these processes.
Only at a concentration of 10,000 ppm these activities were totally
suppressed (23).In soil perfusion studies, employing a garden soil amended
with $CdSO_4$, nitrification was reduced at a concentration 10 mM and almost
completely inhibited by 40 mM (19). During incubation under aerobic conditions of an alluvial sandy loam soil (pH=7.1) only 10,000 ppm of Cu, 1,000
ppm of Cr^{3+} and 100 ppm of Zn depressed ammonification;while nitrification
was also reduced by 100 ppm of Mn. Under anaerobic conditions the effect of
trace elements resulted generally lower.The very low solubility of metals
in the slightly alkaline conditions of this soil, probably explains the
lack of important effects even with the heaviest applications (25).In fact,
in a sandy loam soil with pH=5.8 (38) 288 ppm of Zn were without influence,
while 14.6 ppm or more of Cd produced significantly adverse effects on
bacteria, yeasts, CO_2 evolution and nitrification.Cd>Co>Cu>Zn, from 15 to
195 ppm, depressed symbiotic nitrogen fixation and nodulation of red clover
(22).Nodule weight of soybean, inoculated with Rhizobium japonicum and
cultured in a sand-vermiculite mixture, decreased with all concentrations
of Cd ranging from 18 to 900 µM (14).18 µM inhibited pod fresh weight by
35%. A similar result was produced by 300 µM Pb.

Microbial populations of soil, responsible for the most interesting processes conditioning soil fertility, are therefore very differently sensitive to heavy metals pollution. Evaluation of such a sensitivity remains quite difficult. Duxbury (10) has developed a semi-synthetic medium to determine metal tolerant and non-tolerant bacteria in an australian soil containing metals amounts well within the normal range of concentrations recorded on a global basis. The relative toxicity resulted Hg>Cd>Cu>Ni=Zn, and metal concentrations suitable to differentiate between metal tolerant and non-tolerant soil bacteria were found to be Cd, 0.26mM; Cu, 1.33 mM; Hg, 0.02 mM; Ni, 1.70 mM; Zn, 1,68 mM. Unfortunately the results may be only referred to microorganisms able to grow on the utilized medium and the method was not comparatively applied to polluted soils.

Direct evaluation of every single microbial activity, by appropriate analytical procedure, appears as the unique criterion to ascertain the influences of heavy metals upon microbiological conditions of treated soils. Ammonification, nitrification and non-symbiotic nitrogen fixation were investigated at the University of Naples in two different soils amended with sewage sludge spiced with $CdSO_4$ to have in the treated soils, 0, 2, 4, 8 and 16 ppm of Cd. These soils were also utilized to evaluate the effects of Cd on crop plants by standardized pot trials. Microbiological researches followed the rye grass culture (4 cuts). Pedological characteristics, Cd-bonding capacity, and crop response to Cadmium in the two soils are in the table n.1. Evaluation of activities and microbial counts were performed by previously referred methods (8).

With reference to the groups assayed, the behaviour of soil microorganisms appeared quite different in the two soils. The table 2 reports the MPN of ammonifiers, nitrifiers and nitrogen-fixers xg of control and treated soil. Ammonifier microorganisms resulted strongly inhibited by Cd in the volcanic soil, more weakly in "terra rossa". A similar effect is reflected in the figures relative to mineralization rate of organic nitrogen.

The number of nitrifiers was reduced at a low extent by Cd in volcanic soil, almost unaffected in "terra rossa". However nitrification proceeded without troubles in all the samples during the incubation, and at the end of the experiment, 97÷99% of soil inorganic nitrogen was detected as NO_3^-. Such a result agrees with other esperiences in neutral soils and confirms the ability of some kinds of nitrifiers to tolerate some adverse conditions.

Table 1.　　PEDOLOGICAL CHARACTERISTICS, CADMIUM SOLUBILITY AND RYE GRASS YIELD IN TWO ITALIAN SOILS

	TERRA ROSSA	VOLCANIC
SOIL TYPE AND ORIGIN	TERRA ROSSA from Castellana (Bari), iron-oxides rich (3.6±9%) with high percentage (up to 97%) of non-expansible kaolinite-type clay	VOLCANIC soil developed on the yellow tuff of Posillipo(Naples),fine texured with high percentage of vitreous material and very low clay content
OCCURRENCE	Italy: Apulia, Latium, Venetia Julia, Tuscany Liguria, Sardinia.Southern France, Spain, Greece Yugoslavia, Algeria, Israel	N-NW area of Naples(Campi Flegrei);Isles of Ischia and Procida; volcanic areas of Latium
pH	6.6	6.4
Clay (%)	69.30	7.70
Silt (%)	20.30	16.80
Humus (%)	1.00	1.20
C.E.C. (meq/100 g)	28.88	9.72
Heavy metals (ppm)	Cd=0.15; Zn=134; Cu=11; Mn=959; Fe=34,100	Cd=0.15; Zn=107; Cu=12; Mn=519; Fe=18,600

TERRA ROSSA

Cd (ppm), after treatment with	0.5M NH_4OAc + EDTA	0.1 M $NaNO_3$	Acidification at pH							
			2.5	3.0	3.5	4.0	4.5	5.0	5.5	6.0
0	<0.1	<0.1								
2	1.95	<0.1	1.2	0.8	0.6	0.3	<0.1			
4	3.45	<0.1	1.8	1.2	0.7	0.5	<0.1			
8	7.05	<0.1	3.5	2.5	1.9	0.6	0.4	0.1		
16	11.75	<0.1	7.6	5.5	3.5	2.0	1.6	0.8	0.7	0.6

VOLCANIC

Cd (ppm), after treatment with	0.5M NH_4OAc + EDTA	0.1 M $NaNO_3$	Acidification at pH							
			2.5	3.0	3.5	4.0	4.5	5.0	5.5	6.0
0	0.15	<0.1	<0.1							
2	2.15	<0.1	1.7	1.0	0.7	0.4	<0.1			
4	3.65	0.1	3.6	2.0	1.5	0.8	0.6	0.4	0.3	0.2
8	6.50	0.6	6.0	4.3	2.9	1.7	1.3	0.9	0.8	0.3
16	14.15	1.2	13.3	8.2	6.2	3.3	2.6	2.4	1.9	0.4

CROP RESPONSE
(y=g of plant x pot; x=Cd ppm of soil)

TERRA ROSSA: $y=18.33-2.27x+0.25x^2-0.01x^3$
$r^2=0.97$; F=4.82 (P<0.05)

VOLCANIC: $y=15.76-0.49 \times \ln x$
$r^2=0.94$; F=4.76 (P<0.05)

Cd UPTAKE
(y=μg Cd xg of plant; x=Cd ppm of soil)

TERRA ROSSA: $y=57.45+8.64x-0.63x^2+0.03x^3$
$r^2=1.00$; F=8.72 (P<0.01)

VOLCANIC: $y=3.02+75.85 \times \ln x$
$r^2=0.99$; F=23.43 (P<0.01)

Table 2.   MICROBIAL CONTENTS OF TWO ITALIAN SOILS TREATED WITH DIFFERENT QUANTITIES OF CADMIUM

(Most Probable Number of viable cells $\times$ g^{-1} of soil, d.w.)

Soil	Cd-dose (ppm)	Ammonifiers	NH$_4^+$-oxydizers	NO$_2^-$-oxydizers	Free living Nitrogen fixers	
					Aerobic	Anaerobic
TERRA ROSSA	0	1.03×10^6	4.89×10^3	2.72×10^2	1.63×10^2	4.89×10^2
	2	4.85×10^5	4.02×10^3	2.69×10^2	1.02×10^2	3.23×10^2
	4	8.15×10^4	2.72×10^3	2.72×10^3	1.03×10^2	1.63×10^2
	8	6.50×10^4	2.72×10^2	2.73×10^3	9.73×10^1	1.04×10^2
	16	1.63×10^4	2.71×10^2	2.71×10^3	4.88×10^1	1.03×10^2
VOLCANIC	0	2.59×10^8	2.07×10^4	2.59×10^4	2.59×10^1	9.84×10^3
	2	2.58×10^8	9.64×10^3	9.64×10^3	2.58×10^1	4.64×10^3
	4	4.65×10^7	3.58×10^3	2.58×10^2	2.58×10^1	2.58×10^3
	8	4.63×10^5	3.08×10^3	4.63×10^1	2.57×10^1	9.78×10^2
	16	9.85×10^4	4.67×10^3	2.59×10^1	2.59×10^1	2.59×10^2

EFFECT OF Cd ON AMMONIFICATION
IN A VOLCANIC SOIL

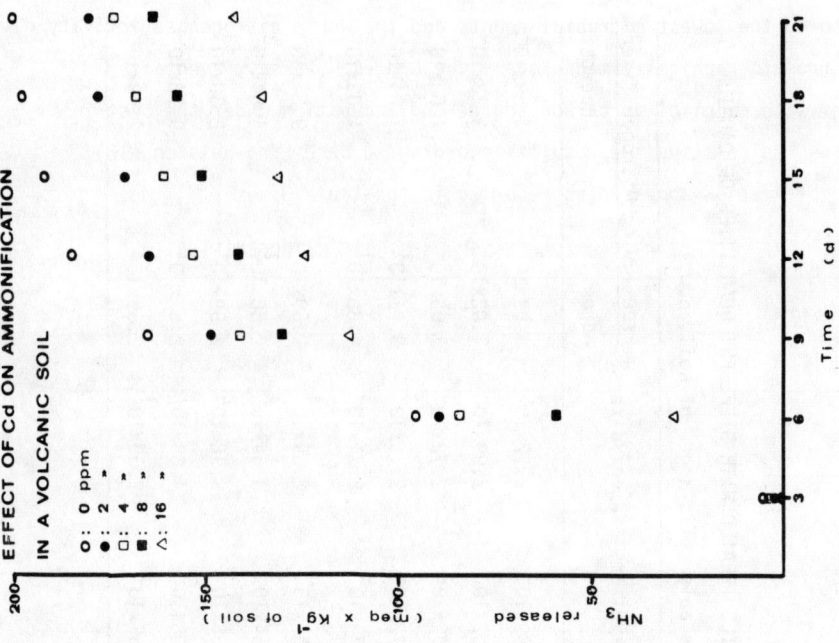

0 ppm
O: 0 ppm
●: 2 "
□: 4 "
■: 8 "
△: 16 "

NH_3 released (meq $\times$ Kg^{-1} of soil)

Time (d)

EFFECT OF Cd ON AMMONIFICATION
IN TERRA ROSSA

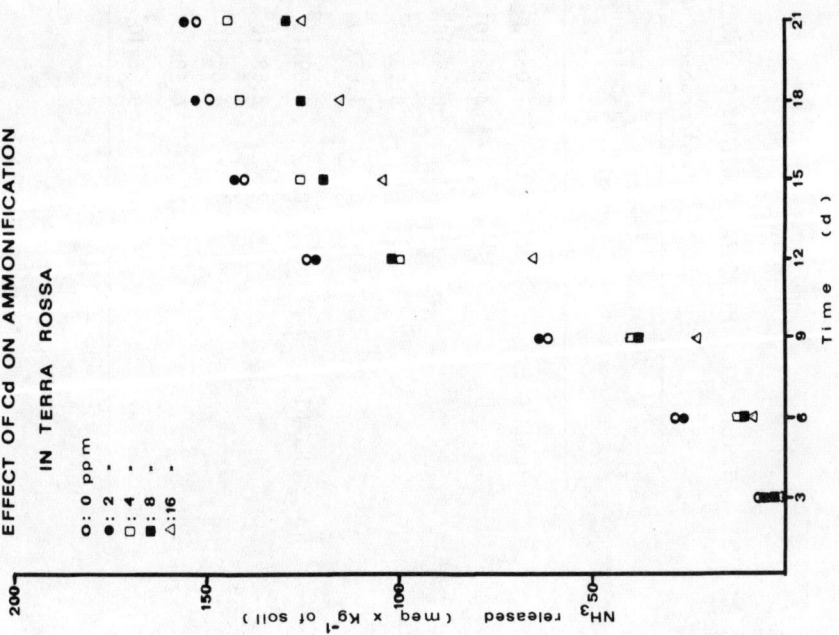

O: 0 ppm
●: 2 "
□: 4 "
■: 8 "
△: 16 "

NH_3 released (meq $\times$ Kg^{-1} of soil)

Time (d)

Both "terra rossa" and volcanic soil showed a poor content of free-living nitrogen fixer microorganisms. Soil samples treated with the highest doses of Cd showed the lowest microbial counts and the worst nitrogenase activity, as gas-chromatographically measured by the C_2H_2-C_2H_4 assay, even after amendment with mannitol as carbon source and incubation under the best conditions. Calculating the results according to Hardy and Holsten (13), dinitrogen fixation appeared as reported in the table 3.

Table 3.   EFFECT OF CADMIUM ON NON-SYMBIOTIC NITROGEN FIXATION

Cd-dose (ppm)	g of N_2 fixed x hectar^{-1} x day^{-1}	
	terra rossa	volcanic soil
0	4,757	2,710
2	2,845	2,566
4	2,228	2,075
8	736	378
16	147	367

Cadmium therefore affects free living nitrogen fixers, but in each soil its influence particularly depends on the biological characteristics of the autochthonous microflora besides on environmental conditions, since in microorganisms metal sensitivity can vary with genera, species, and microbial strains. Hence heavy metals pollution of soil can allow "invisible injuries" too, avoidable by appropriate controls only.

REFERENCES

1) ABELSON,P.H. and E.ALDONS.1950.Ion antagonisms in microorganisms:inter-ference of normal magnesium metabolism by nickel,cobalt,cadmium,zinc and manganese.J.Bacteriol.,60,401-413.
2) BABICH,H. and G.STOTZKY.1977.Sensitivity of various bacteria,including Actinomycetes,and Fungi to Cadmium and the influence of pH on sensitivi ty.Appl.Envron.Microbiol.,33,681-695.
3) BABICH,H. and G.STOTZKY.1977.Reductions in the toxicity of Cadmium to microorganisms by clay minerals.Appl.Environ.Microbiol.,33,696-705.
4) BABICH,H. and G.STOTZKY.1978.Effects of Cadmium on the Biota:influence of environmental factors.In D.PERLMAN (Ed.):Advances in Applied Microbio logy.Vol.23,55-117.Academic Press,New York.
5) BOND,H.,LIGHTHART,R.,SHIMABUKU,R. and L.RUSSEL.1976.Some effects of Cd on coniferous forest soil and litter microcosms.Soil Sc.,121,278-287.
6) CHOPRA,J.1971.Decreased uptake of cadmium by a resistant strain of

Staphylococcus aureus.J.Gen.Microbiol.,63,265-267.

7) CHOPRA,J.1975.Mechanism of plasmid-mediated resistance to cadmium in S. aureus.Antimicrob.Agents Chemother.,7,8-14.

8) COPPOLA,S.1981.Soil microbial activities as affected by applications of composted sewage sludge.EEC Concerted Action "Sewage sludge",Symposium of Working Group 4,München,23th-25th June

9) DRUCKER,H.GARLAND,T.R. and R.E.WILDUNG.1974.in BABICH and STOTZKY (4).

10)DUXBURY,T.1981.Toxicity of heavy metals to soil bacteria.FEMS Microbiol. Letters,11,217-220.

11)GADD,G.M. and A.J.GRIFFITHS.1978.Microorganisms and heavy metal toxicity. Microbial ecology,4,303-317.

12)HAGHIRI,F.1974.Plant uptake of cadmium as influenced by cation exchange capacity,organic matter,zinc and soil temperature.J.Environ.Qual.3,180-182.

13)HARDY,R.W.F. and R.D.HOLSTEN.1977.Methods for measurement of dinitrogen fixation.In HARDY,R.W.F. (GenEd.) and A.H.GIBSON (Section Ed.).A Treatise on dinitrogen fixation.Section IV:Agronomy and Ecology.J.Wiley & Sons, New York.

14)HUANG,C.Y.,BAZZAZ,F.A. and L.N.VANDERHOEF.1974.The inhibition of soybean metabolism by cadmium and lead.Plant Physiol.,54,122-124.

15)JEFFRIES,T.W.1982.The microbiology of Mercury.In M.J.BULL (Ed.):Progress in industrial microbiology.Vol.16,21-118.Elsevier Scientif.Publ.Amsterd.

16)JELLINEK,H. and S.SANGAL.1972.Complexation of metal ions with natural polyelectrolytes (removal and recovery of metal ions from polluted waters).Water Res.,6,305-314.

17)KONDO,I.ISHIKAWA,T. and H.NAKAHARA.1974.Mercury and cadmium resistances mediated by the penicillinase plasmid in S.aureus. J.Bacteriol.117,1-4.

18)LAWRENCE,A.W. and P.L.McCARTY.1965.The role of sulphide in preventing metal toxicity in anaerobic treatment.J.Water Pollut.Contr.Fed.,37,392-406.

19)LEES,H. and J.MEIKLEJOHN.1948.Trace elements and nitrification.Nature (London),161,398-399.

20)LEVI-MINZI,R.,SOLDATINI,G.F. and R.RIFFALDI.1976.Cadmium absorption by soils.J.Soil Sci.,27,10-15.

21)MALISZEWSKA,W.1972.Influence de certains oligo-éléments sur l'activité de quelques processus microbiologiques du sol.Rev.Ecol.Biol.Sol.9,505-512.

22)McILVEEN,W.D. and H.COLE Jr.1974.Influence of heavy metals on nodulation of red clover.Phytopathology,64,583.

23)MORISSEY,R.F.,DUGAN,E.P. and J.S.KOTHS.1974.In BABICH and STOTZKY (4).

24)PICKETT,A.W. and A.C.R.DEAN.1976.Cadmium and zinc sensitivity and tolerance in Klebsiella (Aerobacter) aerogenes.Microbiology,15,79-91.

25)PREMI,P.P.and A.H.CORNFIELD.1969.Effects of addition of copper,manganese, zinc and chromium compounds on ammonification and nitrification during incubation of soil.Pl.Soil,31,345-352.

26)RIDLEY,W.P.,DIZIKIES,L.J. and J.M.WOOD.1977.Biomethylation of toxic elements in the environment.Science,197,329-332.

27)ROSENZWEIG,W.D. and D.PRAMER.1980.Influence of cadmium,zinc and lead on growth,trap formation,and collagenase activity of nematode-trapping fungi.Appl.Environ.Microbiol.,40,694-696.

28)RUHLING,A. and G.TYLER.1973.Heavy metal pollution and decomposition of spruce needle litter.Oikos,24,402-416.

29)SADLER,W.R.and P.A.TRUDINGER.1967.The inhibition of microorganisms by heavy metals.Mineral Dep.,2,158-168.

30)SAXENA,J. and P.H.HOWARD.1977.Environmental transformation of alkylated and inorganic forms of certain metals.Adv.Appl.Microbiol.,21,136-158.

31)SILVER,S.,SCHOTTEL,J. and A.WEISS.1976.Bacterial resistance to toxic metals determined by extrachromosomal R-factors.In J.M.Sharpley and A.M. Kaplan (Eds.):Proceedings of the third international biodegradation Symposium.Appl.Sci.Publishers, London.

32)STEVENSON,F.J.1976.Binding of metal ions by humic acids.In J.O.Nriagu (Ed.):Environmental Biogeochemistry.Ann Arbor Sci.,Ann Arbor, Mich.

33)STUTZENBERG,F.J. and E.O.BENNETT.1965.Sensitivity of mixed populations of Staphylococcus aureus and Escherichia coli to mercurials. Appl. Microbiol.,13,570-574.

34)TYLER,G.1974.Heavy metal pollution and soil enzymatic activity.Pl.Soil, 40,303-311.

35)TYLER,G.1975.Heavy metal pollution and mineralization of nitrogen in forest soils.Nature (London),255,701-702.

36)TYLER,G.1976.Heavy metal pollution, phosphatase activity, and minerali-zation of organic phosphorus in forest soils.Soil Biol.Biochem.,8, 327-332.

37)TYLER,G.,MORNSJO,B. and B.NILSSON.1974.Effects of cadmium,lead,and sodium salts on nitrification in a mull soil.Pl.Soil,40,237-242.

38)WALTER,C. and F.STADELMANN.1979.Influence du zinc et du cadmium sur les microorganismes ainsi que sur quelques processus biochimiques du sol. Schweiz.landw.Forschung,18,311-324.

39)WOOD,J.M.1974.Biological cycles for toxic elements in the environment. Science,183,1049-1052.

40)YOUNG,R.G.and D.J.LISK.1972.Effect of copper and silver ions on algae J.Water Pollut.Control Fed.,44,1643-1647.

41)ZIMMERMAN,L.1966.Toxicity of copper and ascorbic acid to Serratia marcescens. J.Bacteriol.,91,1537-1542.

HEAVY METALS IN SOILS, SLUDGES AND PLANTS : SUMMARY OF
RESEARCH ACTIVITIES

A. COTTENIE & L. KIEKENS
Laboratory of Analytical and Agrochemistry
State University Ghent - Belgium

In the past quite a lot of research has been done on phytotoxic effects of heavy metals (pot experiments), and their behaviour in soils. The effect of some soil parameters (pH, CEC, organic matter) on heavy metal availability and uptake by plants has been studied. Chemical analysis of a number of sludges has been carried out whereby as well total metal concentrations as different metal fractions (selective extractants) were determined (see publications).

At present a procedure is being developed for the speciation of heavy metals in soils and sludges. The methodology is primarily based on charge separation.

A number of equilibration and pot experiments are being carried out with the following objectives :
- evolution in function of time of heavy metal availability in sludge and sludge amended soil
- uptake and phytotoxic effects of heavy metals by a number of indicator plants. Comparison between the effects of the metals when present under different forms : inorganic, chelated, present in sludge
- determination of the coefficients of transfer of the different heavy metals
- evolution of the different metal species in function of time
- influence of organic matter breakdown on heavy metal availability.

The possibilities of reducing heavy metal availability in contaminated soils is studied and the relative importance of precipitation (pH increase), absorption (increase of CEC), complexation (increase of organic content) is being evaluated.

A study of the importance of atmospheric fall-out with regard to heavy metals in industrial and rural areas is being carried out.

A survey of the trace element content of cultivated soils and plant species in a region of about 500 km^2 is under study.

List of publications

1. COTTENIE, A. & KIEKENS, L. 1974. Quantitative and qualitative plant response to extreme nutritional conditions. Proc. 7th IPI Colloq., Hannover.

2. COTTENIE, A., KIEKENS, L. & VERLOO, M. 1975. Principles of soil and substrate analysis with regard to mobility of nutrient elements. Pedologie, XXV, 134-142.

3. COTTENIE, A., DHAESE, A. & CAMERLYNCK, R. 1976. Plant quality response to uptake of polluting elements. Qualitas Plantarum - Pl. Fds. Hum. Nutr. XXVI, 1/3, 293-319.

4. COTTENIE, A., DHAESE, A. 1978. Caractérisation chimique des boues. Annales de Gembloux, n° 84, pp. 25-34.

5. COTTENIE, A., DHAESE, A. 1978. Content and activity of heavy metals in soils, sediments and water. Med. Fac. Landbouww. Rijksuniversiteit Gent 43.

6. COTTENIE, A., VERLOO, M., KIEKENS, L., CAMERLYNCK, R., VELGHE, G., DHAESE, A. 1979. Essential and non essential trace elements in the system soil-water-plant. Laboratory of Analytical and Agrochemistry, State University Ghent, Belgium.

7. COTTENIE, A., CAMERLYNCK, R., VERLOO, M., DHAESE, A. 1979. Fractionation and determination of trace elements in plants, soils and sediments. Pure & Appl. Chem., Vol. 52, pp. 45-53.

8. COTTENIE, A., KIEKENS, L. 1980. Beweglichkeit von Schwermetallen in mit Schlamm angereicherten Böden. Korrespondenz Abwasser, 28 Jg, n° 4, 206-210.

9. COTTENIE, A. et al. 1981. Trace elements in agriculture and in the environment. Laboratory of Analytical and Agrochemistry, State University Ghent, Belgium.

10. COTTENIE, A., VELGHE, G, VERLOO, M., KIEKENS, L. 1982. Biological and analytical aspects of soil pollution. Laboratory of Analytical and Agrochemistry, State University Ghent, Belgium.

11. DHAESE, A., COTTENIE, A. 1979. Contents of heavy metals in sludges and their environmental significance. CEC Proceedings Symposium "Treatment and use of sewage sludge", Cadarache 12-15/2/79 (Ed. J. Alexandre & H. Ott).

12. KIEKENS, L., VERLOO, M., COTTENIE, A. 1980. Uptake of zinc by corn as influenced by clay and humic acid content of the substrate. 5e Coll. Intern. sur le Contrôle de l'Alimentation des Plantes Cultivées, 25-30.8.1980, Castelfranco, Veneto (Italy).

DISCUSSION ON SESSION II: EFFECTS ON INORGANIC MICROPOLLUTANTS

Chairman: R.D. DAVIES - Rapporteur: J.E. HALL

(After papers by Tjell, and Sherlock)

G Hucker (UK)

What is the soil limit beyond which proportionality between soil and plant contents of cadmium is lost.

J Tjell (Denmark)

There is proportionality up to perhpas 10-15 ppm of cadmium in soil. But I think we should concern ourselves with much lower levels than this.

P Beckett (UK)

The intercept on your graph of soil and plant concentrations is on the y-axis. Is this native cadmium not available?

J Tjell

The explanation for this is aerial deposition of cadmium.

N Nicolson (UK)

May I make the observation that the level of cadmium at 3-4 ppm (dry solids) in Danish sludge is very low. In London we never get below 10 ppm of cadmium. But my question concerns the availability of cadmium. Is cadmium in inorganic fertilisers as available as cadmium in sludge?

J Tjell

The source of cadmium is of little importance. After 2-3 years the availability in the soil is the same.

H Kuntze (Germany)

Dr Tjell has shown a correlation between soil texture and pH value with cadmium uptake. We have found that uptake from poorly aerated clay soils of low pH value can be as great as from acid sands. Has Dr Tjell observed this effect?

J Tjell

There are no soils of high clay content in Denmark.

D Sauerbeck (Germany)

I agree with most of Dr Tjells' comments but not his generalisation on fertiliser and sewage sludge availabilities. We have evidence that cadmium in mine spoil is more available than from sewage sludge compost.

After papers by Hani, Webber and Davis:

A Kloke (Germany)

Methods are required to determine how much metal in the soil gets into the plant. Uptake is dependent not only on the plant but also on climatic conditions and changes in soil pH value due to the use of inorganic fertiliser. An experiment started 20 years ago with inorganic fertiliser has caused the soil pH value to drop from 6.5 to 4.9 and the cadmium content of plants increased because of this even though no cadmium had been added to the soil. Metal limits in the soil must apply to total not extractable metal to allow for future changes in soil conditions which cannot be predicted.

H Hani (Switzerland)

I accept the importance of total concentrations of metal in soil but concentrations of soluble metal are of more direct relevance to plant uptake. Increases in soluble amounts of metal can be followed as the soil pH value changes.

D Sauerbeck

I agree with Dr Hani except that I am not entirely happy with sodium nitrate as an extractant.

S Berglund (Sweden)

I do not agree with the level of 5 kg of cadmium per hectare suggested by Dr Webber as being acceptable for sludge-treated soil. Has he considered long-term effects such as changes in soil humus content and pH value and perhaps changes in land use such as growing sensitive vegetable crops instead of grass. Other unknown factors in the future include the aerial deposition of acid compounds, the use of inorganic fertilsers, and deposition of airborne cadmium which may continue after the 5 kg/ha limit has been reached. An addition of 5 kg/ha of cadmium would mean a tenfold rise in the cadmium content of top soil (20 cm) in Sweden. We cannot accept such a rise.

M Webber

I agree that adding 5 kg Cd/ha might increase the cadmium content of Swedish soils tenfold. The same would be true for many Canadian soils. However, increasing evidence, including the United States diet scenario risk analysis, indicates that this level of addition presents a very low hazard to human health even where garden vegetables are grown on soils acidified by fertiliser use, aerial deposition of acidic materials and soil weathering. Soil pH value is the most important soil property affecting cadmium uptake. The effects of soil organic matter content and cation exchange capacity are of secondary importance. My recommendations did not take into account aerial deposition of cadmium to soil.

After papers by Sauerbeck and Gomez:

P Beckett

In the literature, available metal is often referred to. Professor
Sauerbeck's work shows that when chemical extractants are used we should speak
of extractable metal not available metal. Professor Sauerbeck used the term
'complexly bound metal'. What exactly does this mean?

D Sauerbeck

I cannot easily answer this. It is true that metal extractable by a complexing agent is not necessarily complexly bound.

P Beckett

You produced some suggestions as to which extractants are most useful.
Were these conclusions supported by plant trials of the quantities extracted?

D Sauerbeck

There was limited evidence of correlation. The idea was merely to screen
extractants.

L Kiekens (Belgium)

Professor Sauerbeck used copper acetate in his fractionation for the
determination of complexed zinc, cadmium and lead. Which extractant does he
use for copper which is an element known to form stable complexes with soil
organic constituents? I also observe that in your mobility procedure you use
0.5 N nitric acid to adjust the pH value of the soil suspension. Our experience
has shown that in some cases stronger concentrations of nitric acid (up to
4.0 N) have to be used to adjust the pH level to low values.

D Sauerbeck

I have no precise answer. We have not considered copper yet.

J Tjell

I would like to ask Dr Gomez whether cadmium is usually bound organically
or inorganically in normal soils. He showed it bound only to humates.

A Gomez (France)

The soil we used was an organic sand in a maize growing area. In this
soil binding of cadmium to organic matter is the principal mechanism. In these
circumstances the activity of soil micro-organisms can be important in controll
ing the solubility of cadmium.

D Sauerbeck

Did Dr Gomez purify the cadmium solutions and measure dissolved cadmium
ions.

A Gomez

We measured total cadmium in solution after filtering out organic matter.

After papers by Beckett, and Williams:

Chairman

Dr Beckett's data show decreasing availability of copper and zinc over the four years after the application of sludge. Has the soil been limed during this time.

P Beckett

It was limed originally to pH value 6.5 and the pH has been monitored since. There is evidence for a decline in availability.

P Worthington (UK)

The loss in yield due to "lodging" of the crop reported by Dr Beckett can often be reduced considerably by the use of manual harvesting techniques rather than machine harvesting. Was this approach considered by Dr Beckett?

P Beckett

The main difficulty is that "lodging" encourages fresh growth.

G Fleming (Ireland)

Would Dr Beckett like to comment further on metals levels in soils in long-term experiments? I have found a gradual disappearance in "total" metal levels by time-dependent adsorption into clay lattices.

P Beckett

We have observed that "native" forms of metal in soil are less available for plant uptake than metal introduced to the soil in sludge. The use of "total" metal analysis ignores the fact that some metal will revert to "native" forms as time passes.

J Tjell

I think this is true except for cadmium and lead which remain unreverted for long periods. Metals may vary in the time taken to revert.

D Sauerbeck

Did Dr Beckett really expect to see toxicity at his highest treatment which added to the soil 1600 kg/ha of zinc. In Germany applications of 3000 kg Zn/ha in some experiments had not caused toxicity although they increased crop levels of zinc.

P Beckett

Rate of metal addition at the top rate was about four times the maximum recommended level in UK guidelines.

Chairman

The application of 1600 kg/ha of zinc would produce a soil concentration of about 800 ppm of zinc compared with Mr Williams proposed safe limit of 275 ppm. In this sense it seemed a potentially toxic application rate.

After papers by Kloke, and Berglund:

D Sauerbeck

I find Mr Berglund's figures on cadmium in the kidney are rather alarming. They do not agree with existing evidence. Can he confirm that these findings are correct.

J Sherlock (UK)

The critical concentration of cadmium in the kidney cortex identified by the World Health Organisation (WHO) is 200 μg/g. More recent evidence suggests that the critical concentration may be even higher. In addition, the currently accepted absorption coefficient and half-life for cadmium in man are thought to be pessimistic. Therefore the WHO Provisional Tolerable Weekly Intake for cadmium (400-500 μg) could well be increased when cadmium is considered next by the FAO/WHO.

S Berglund

I accept that there are considerable differences of opinion here.

J Tjell

Other evidence suggests 200 μg/g of cadmium in the kidney is too high. I think we do not yet know the right answer.

A Kloke

We must not ignore the risks from cadmium. We must work to minimise cadmium levels in our food.

J Sherlock

I agree that the risks must not be ignored. But if we are really concerned about cadmium then smoking cigarettes is one of the worst risks.

Chairman

It is interesting to notice that whilst there is uncertainty about the toxic effects of cadmium there is reasonable agreement on a soil concentration limit for cadmium at around 3 ppm. For instance, Professor Kloke's figure is 3 ppm and in current UK guidelines it is 3.5 ppm. Mr Berglund's paper has demonstrated highest cadmium availability to crops in the early years after application of sludge to soil with a decline in availability thereafter. This is an important aspect. We need to know the trend concerning long-term availability of metals added to the soil in sludge. More evidence is needed from long-term field trials.

After papers by Alloway, and Herms:-

J Tjell

A comparatively small amount of metal is taken up by plants following applications of sludge to soil. Does Mr Herms think this is due to redox conditions in the soil.

<u>U Herms (Germany)</u>

There would be little mobilisation with only slightly reducing conditions in the soil.

<u>P Worthington</u>

Since there was an indication in Dr Alloway's work that cadmium existed in sewage sludge predominantly as a single ionic species, had he calculated solubility products of the ions in water and checked whether these were in agreement with the soil solution concentrations.

<u>B Alloway (UK)</u>

It has proved difficult to determine accurately the low levels of cadmium in soil solution. The work was mainly qualitative.

<u>Chairman</u>

Dr Alloway's results were interesting because it was surprising that cadmium from such different soils was so consistent in form in soil solution.

After papers by Kiekens, Fleming and Coppola:

<u>A Kloke</u>

Our work with selenium and molybdenum in Berlin has shown equal concentrations of selenium in grain and straw but much higher concentrations of molybdenum in straw.

<u>G Fleming</u>

We must be very careful in interpreting data from natural and contrived conditions.

<u>D Sauerbeck</u>

We have seen no yield decreases in legumes at up to >200 ppm of molybdenum in the soil.

<u>G Fleming</u>

It must be remembered that molybdenum and selenium are animal nutrition problems not phytotoxic elements.

<u>B Alloway</u>

We have found Namurian derived soils rich in cadmium, lead and arsenic.

<u>G Fleming</u>

Cadmium levels were not as high as expected in the Irish work but we will keep looking.

<u>M Webber</u>

Has there been any effect on the people living in these contaminated areas.

<u>G Fleming</u>

During the last century when potatoes were the staple diet , there was a a legend of loss of hair and nails but there is nothing in the literature to support this. Work in the USA is showing effects on human health.

J Tjell

The results of Professor Coppola showed that 2 ppm of cadmium in the soil was enough to reduce nitrogen fixation by soil bacteria. This was a disturbingly low level bearing in mind that soil limits for cadmium were set at above 2 ppm in many countries.

S Coppola (Italy)

Non-symbiotic nitrogen fixation was very sensitive to cadmium pollution. Effects on soil microbes as well as on crops should be considered in the setting of limits.

Chairman

It was unusual to see from Professor Coppola's results that the more toxic effects were seen on the soil with the highest pH value and cation exchange capacity. Usually, the opposite occurred. The last two papers, in dealing on the one hand with the black shales of Ireland and on the other with volcanic soils of Southern Italy, had exemplified the wide range of soil conditions to be found within Europe. Was it adequate to have soil metal limits which took no account of soil conditions beyond controls on pH value?

LIST OF PARTICIPANTS

AHTINAINEN, M.
  National Board of Waters
  Water District Office of
  North Karelia
  Torikatu 36A
  SF - 80101 JOENSUU 10

ALLOWAY, B.J.
  Westfield College
  University of London
  Kipperpore Avenue
  UK - LONDON NW3 7ST

BECKETT, P.H.T.
  Oxford University
  Dept. of Agricultural Science
  Parks Road
  UK - OXFORD

BERGLUND, S.
  Swedish Environment
  Protection Board
  P.O. Box 1302
  S - 17125 SOLNA

BRIDLE, T.
  Environment Canada
  Wastewater Technology Center
  Box 5050
  Canada - BURLINGTON,
           ONTARIO L7R 4A6

CALCUTT, T.
  Water Research Centre
  Process Engineering
  Stevenage Laboratory
  Elder Way
  UK - STEVENAGE SG1 1TH, Herts.

COPPOLA, S.
  University of Naples
  Institute of Agricultural
  Microbiology
  I - 80055 PORTICI

DAVIS, R.D.
  Water Research Centre
  Elder Way
  UK - STEVENAGE SG1 1TH, Herts.

DIERCXSENS, Ph.
  Ecole Polytechnique Fédérale
  de Lausanne
  Institut du Génie de
  l'Environnement
  88, Bredabaan
  B - 2130 BRASSCHAAT

DUVOORT-VAN ENGERS, L.E.
  Instituut voor
  Afvalstoffenonderzoek
  Natriumweg 7
  NL - 3800 AD AMERSFOORT

FLEMING, A.G.
  An Foras Taluntais
  The Agricultural Institute
  Johnstown Castle Research Centre
  Ireland - WEXFORD

GOMEZ, A.
  Institut National de la
  Recherche Agronomique
  Station d'Agronomie
  La Grande Ferrade
  F - 33640 PONT DE LA MAYE

GUPTA, S.K.
  Swiss Federal Research Station
  for Agricultural Chemistry and
  Hygiene of Environment
  Schwarzenburgstr. 155
  CH - 3097 LIEBEFELD-BERN

HALL, J.
  Water Research Centre
  Stevenage Laboratory
  Elder Way
  UK - STEVENAGE SG1 1TH, Herts.

HAENI, H.
  Swiss Federal Research Station
  for Agricultural Chemistry and
  Hygiene of Environment
  Schwarzenburgstr. 155
  CH - 3097 LIEBEFELD-BERN

HERMS, U.
Niedersächsisches Landesamt fuer
Bodenforschung - Bodentechno-
logisches Institut Bremen
Friedrich-Missler-Str. 46-50
D - 2800 BREMEN 1

HUCKER, G.
Department of the Environment
Romney House
43 Marsham Street
UK - LONDON SW1

KIEKENS, L.
State University Gent
Faculty of Agricultural Sciences
Coupure Links 653
B - 9000 GENT

KLOKE, A.
Federal Biological Research Centre
for Agriculture and Forestry
Koenigin-Luise-Str. 19
D - 1000 BERLIN 33

KUNTZE, H.
Niedersächsisches Landesamt fuer
Bodenforschung - Bodentechno-
logisches Institut
Friedrich-Missler-Str. 46-50
D - 2800 BREMEN 1

LESCHBER, R.
Institut für Wasser-, Boden- und
Lufthygiene des Bundesgesundheits-
amtes
Corrensplatz 1
D - 1000 BERLIN 33

LESTER, J.
Imperial College
Imperial College Road
UK - LONDON SW7 2BU

L'HERMITE, P.
Commission of the European
Communities
Directorate General "Science,
Research and Development"
200, rue de la Loi
B - 1049 BRUSSELS

LINDSAY, D.G.
Food Science Division
Ministry of Agriculture,
Fisheries and Food
Horseferry Road
UK - LONDON SW1P 2AE

MILLER, D.
Water Research Centre
Medmenham Laboratory
P.O. Box 16 (Medmenham)
UK - MARLOW, Bucks. SL7 2HD

MOSS, J.
Water Research Centre
Medmenham Laboratory
P.O. Box 16 (Medmenham)
UK - MARLOW, Bucks. S17 2HD

ORNONWSKI, E.
Instituut voor Hygiene
en Epidemiologie
Min. van Volksgezondheid
J. Wytsmanstraat 14
B - 1050 BRUSSEL

PORTEOUS, G.
Department of the Environment
Romney House
43 Marsham Street
UK - LONDON SW1

ROCHER, M.
Faculté des Sciences Agronomiques
de l'Etat à Gembloux
B - 5800 GEMBLOUX

SAUERBECK, D.
Federal Agricultural Research
Center Braunschweig-Volkenrode
50 Bundesallee
D - 3300 BRAUNSCHWEIG

SHERLOCK, J.Ch.
Ministry of Agriculture,
Fisheries and Food
Food Science Division
Horseferry Road
UK - LONDON SW1P 2AE

STARK, J.H.
  Water Research Centre
  Stevenage Laboratory
  Elder Way
  UK - STEVENAGE SG1 1TH, Herts.

TILLS, A.R.
  Westfield College, Univ. of London
  Department of Botany and Bio-
  chemistry
  Kipperpore Avenue
  UK - LONDON NW3 7ST

TJELL, J. Chr.
  Technical University of Denmark
  Dept. Sanitary Engineering
  Building 115
  DK - 2800 LYNGBY

VAN DE MAELE, F.
  OVAM (Openbare Afvalstoffen-
  maatschappij voor het Vlaamse
  Gewest)
  Nekkerspoelstraat 21-23
  B - 1800 MECHELEN

VIGERUST, E.
  Agricultural University
  of Norway
  Ekornu 18
  N - 1430 AS

WEBBER, M.D.
  Wastewater Technology Centre
  Environmental Protection Service
  Box 5050
  Canada - BURLINGTON,
              ONTARIO L7R 4A6

WILLIAMS, J.H.
  ADAS Science Service
  Ministry of Agriculture
  Woodthorne
  UK - WOLVERHAMPTON, WV6 8TQ

WORTHINGTON, P.
  Department of Environment
  Romney House, Room A414
  43 Marsham Street
  UK - LONDON SW1P 3PY

INDEX OF AUTHORS